Chemie
67,-€
1/19

Lecture Notes in Management and Industrial Engineering

Series editor

Adolfo López-Paredes, INSISOC, University of Valladolid, Valladolid, Spain

This book series provides a means for the dissemination of current theoretical and applied research in the areas of Industrial Engineering & Engineering Management. The latest methodological and computational advances that both researchers and practitioners can widely apply to solve new and classical problems in industries and organizations constitute a growing source of publications written for and by our readership.

The aim of this book series is to facilitate the dissemination of current research in the following topics:

- Strategy and Enterpreneurship;
- Operations Research, Modelling and Simulation;
- Logistics, Production and Information Systems;
- Quality Management;
- Product Management;
- Sustainability and Ecoefficiency;
- Industrial Marketing and Consumer Behavior;
- Knowledge and Project Management;
- Risk Management;
- Service Systems;
- Healthcare Management;
- Human Factors and Ergonomics;
- Emergencies and Disaster Management;
- Education

More information about this series at http://www.springer.com/series/11786

Raymond R. Tan · Kathleen B. Aviso
Michael Angelo B. Promentilla
Krista Danielle S. Yu · Joost R. Santos

Input–Output Models for Sustainable Industrial Systems

Implementation Using LINGO

De La Salle University

Springer

Raymond R. Tan
Chemical Engineering Department
De La Salle University
Manila, Philippines

Kathleen B. Aviso
Chemical Engineering Department
De La Salle University
Manila, Philippines

Michael Angelo B. Promentilla
Chemical Engineering Department
De La Salle University
Manila, Philippines

Krista Danielle S. Yu
School of Economics
De La Salle University
Manila, Philippines

Joost R. Santos
Department of Engineering Management
and Systems
George Washington University
Washington, DC, USA

ISSN 2198-0772 ISSN 2198-0780 (electronic)
Lecture Notes in Management and Industrial Engineering
ISBN 978-981-13-1872-6 ISBN 978-981-13-1873-3 (eBook)
https://doi.org/10.1007/978-981-13-1873-3

Jointly Published with De La Salle University Publishing House, Manila, Philippines

The print edition is not for sale in Philippines. Customers from Philippines please order the print book from: De La Salle University Publishing House.

Library of Congress Control Number: 2018949889

This Springer imprint is published by the registered company Springer Nature Singapore Pte Ltd.
The registered company address is: 152 Beach Road, #21-01/04 Gateway East, Singapore 189721, Singapore

Preface

Input–output analysis is a powerful and mathematically elegant framework for describing the networks of interdependent components. While originally developed for the analysis and planning of economic systems, over the past few decades, it has also been shown to apply to diverse types of systems. Its ability to reflect the interdependencies among system components is its most essential feature, as it allows input–output models to gauge ripple effects that may cause unintended, and potentially adverse, consequences if not predicted properly.

Sustainability considerations are now widely considered as integral to decision-making in business, engineering, and public policy. This book is intended to provide a concise treatment of different extensions of the input–output analysis as they apply to the planning of sustainable industrial systems. The first two chapters of the book give an introduction to the conceptual and mathematical foundations of the input–output analysis. A unique feature is the use of the commercial optimization software, LINGO, whose equation-based interface is ideal for learning the input–output models. Thus, the third chapter is dedicated entirely to a description of this programming environment. The next six chapters describe different extensions of the input–output model for various industrial systems at different scales, including industrial complexes, infrastructure systems, and supply chains. Further extensions delve into the use of the framework for plant design, operational optimization, human resource planning, and life cycle assessment. All these chapters provide ample illustrative case studies, accompanied by LINGO code to allow the reader to duplicate the results and even to use these models as starting points for their own case studies. In the final chapter of the book, we discuss some research prospects that we consider to be important directions for the future input–output analysis research.

Some acknowledgements are due here. First, we appreciate the cooperation of LINDO Systems, www.lindo.com, in the preparation of this book; in particular, Mark Wiley of LINDO Systems has been especially supportive in responding to our queries. Some of the materials in this book were developed with the support from the Philippine Commission on Higher Education (CHED) via the Philippine Higher Education Research Network (PHERNet) Sustainability Studies Program spanning the period from 2013 to the present. We also want to acknowledge Sean Francis Hinolan for assistance in the preparation of the figures and are grateful for the arrangements made by the editorial teams at Springer Nature and the De La Salle University Publishing House. Finally, we want to thank our respective families for their support and hence dedicate this book to them.

Manila, Philippines Raymond R. Tan
Manila, Philippines Kathleen B. Aviso
Manila, Philippines Michael Angelo B. Promentilla
Manila, Philippines Krista Danielle S. Yu
Washington, USA Joost R. Santos

Contents

Acronyms

ADB	Asian Development Bank
AF	Annualizing factor
AHP	Analytic hierarchy process
ANNs	Artificial neural networks
CE	Circular economy
CHP	Combined heat and power
CIS	Critical infrastructure systems
CRF	Capital recovery factor
DEMATEL	Decision-Making Trial and Evaluation Laboratory
EIP	Eco-industrial park
EU	European Union
FLP	Fuzzy linear programming
GDP	Gross domestic product
IE	Industrial ecology
IIM	Inoperability input–output modeling
I–O	Input–output
IS	Industrial symbiosis
ISO	International Standards Organization
LCA	Life cycle assessment
LCI	Life cycle inventory analysis
LCIA	Life cycle impact assessment
LP	Linear programming
MILP	Mixed-integer linear programming
MOLP	Multiple-objective linear programming
OECD	Organisation for Economic Co-operation and Development
PNS	Process network synthesis
RDUs	Resource depletion units
REPA	Resource and environmental profile analysis
SAW	Simple additive weighting
SETAC	Society of Environmental Toxicology and Chemistry

Chapter 1
Introduction to Input–Output Models

Abstract This chapter provides a general introduction to input–output analysis and input–output models. A brief description of the historical development of the framework, leading to its widespread use, is given. A qualitative discussion of the general framework is presented, followed by a discussion of the key assumptions that underlie input–output models as well as the resultant limitations. The chapter also provides an overview of the rest of the book.

Keywords Input–output analysis · Leontief inverse · Economics

1.1 Introduction

Professor Wassily Leontief was awarded the Nobel Prize in Economic Sciences in 1973 "for the development of the input–output method and for its application to important economic problems.[1]" While its origins go way back in the 1700s when Francois Quesnay proposed an accounting of economic transactions using the "Tableau economique," the input–output approach proved its novelty through presenting a framework for empirical analysis, which was demonstrated empirically via a comprehensive analysis of the economic structure of the USA [10].

Input–output analysis has been widely used in analyzing the interdependencies between economic sectors by representing the economy as a system of linear equations. It is now widely used in many countries as a means to keep track of national accounts. The widespread use of this method can also be seen by the scale of the scientific literature on the topic. For example, searching for "input–output analysis" in Google Scholar yields over 67,000 documents; this figure includes scientific articles, conference papers, books, chapters, and online gray literature. This broad acceptance of input–output analysis is due largely to the elegance and versatility of this framework. Figure 1.1 illustrates the interactions between sectors taking resource inputs and transforming them into final output of each sector for

[1]Nobelprize.org, 'Wassily Leontief—Biographical,' https://www.nobelprize.org/nobel_prizes/economic-sciences/laureates/1973/leontief-bio.html (January 25, 2018).

R. R. Tan et al., *Input–Output Models for Sustainable Industrial Systems*,
Lecture Notes in Management and Industrial Engineering,
https://doi.org/10.1007/978-981-13-1873-3_1

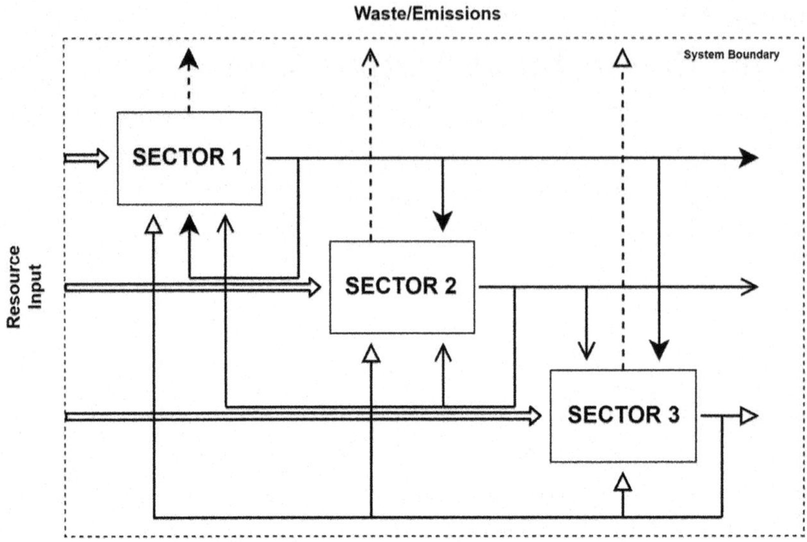

Fig. 1.1 Conceptual framework of an input–output system

both intermediate consumption and final consumption. The economic system is assumed to be balanced or in equilibrium, such that the total inputs to the entire economy add up to the total output.

In addition to the traditional input–output system where resource inputs and final output are accounted for, wastes and emissions resulting from the production process can also be included. Such extensions were first developed in the early 1970s [2]. The usage of natural resources and the generation of wastes and emissions are negative externalities that are not included in the conventional accounting process for economic systems. These streams emanate directly from, or terminate directly to, the external natural environment rather than economic sectors within the system. Accounting for such flows plays an integral role in measuring the sustainability of the production process in industrial systems and identifying opportunities for improvement.

Table 1.1 presents the flow of transactions within a three-sector economic system expressed in monetary terms.[2] The first three rows define economic sectors as producers of output, while the first three columns define economic sectors as consumers of other sectors' output. This block contains information on the interindustry flow of transactions wherein sectors purchase other sectors' output as an input for their production process. In addition to the raw materials, economic sectors also require labor and capital inputs, as well as remit indirect taxes, to process the intermediate inputs. This is presented as the rows under value added for each sector. The sum of the value of the raw materials from the various sectors,

[2]Transaction flows may also be expressed in physical terms as in Hubacek and Giljum [9].

labor and capital inputs, and indirect taxes yields the value of all inputs to each sector. Furthermore, the rows reflect the value of demand for goods and services produced by each economic sector, which sums up to total output. Values under the consuming sectors column show the demand for goods and services used for further processing, and the columns under final demand show the demand for goods and services by consumers, firms, government, and the rest of the world for consumption as end users. Table 1.1 results in a balanced table, wherein the entries for total output are equal to the entries for total input of each sector.

In summary, the input–output table provides a snapshot of the macroeconomic transactions within an economy.

1.2 Assumptions of the Input–Output Model

The input–output model proposes a framework for an economy under certain assumptions such as:

1. Each sector follows the same production process and produces only one good.
2. Sectors use inputs in fixed proportions.
3. There is no supply constraint.
4. The model does not account for changes in price.

These assumptions are discussed in detail in this section.

Each sector follows the same production process and produces only one good.

An input–output table may have varying resolutions depending on the way sectors are disaggregated. For example, the manufacturing sector may be disaggregated into food manufactures, beverage industries, furniture and fixtures, chemical and chemical products, basic metal industries, etc. It can be noted that the sectors may use different production processes; however, upon aggregation into a homogenous sector, it is assumed that they all use a similar representative, or "average," process.

This assumption is justified through introducing flows in monetary terms. By converting the units into a single currency value, the interindustry flows can easily be added and the result is a table with values across sectors that are comparable to one another. Hence, aggregating the subsectors of manufacturing into a single manufacturing sector is possible. In practice, the choice of the level of disaggregation in input–output models depends on conflicting considerations. A higher level of resolution (more disaggregation) allows for more nuanced, high-fidelity modeling; however, it also creates more demand on data collection and verification.

Sectors use inputs in fixed proportions.

Assembling a typical computer station requires a monitor, a central processing unit, a keyboard, and a mouse. Suppose one has a monitor, a central processing unit, two keyboards, and a mouse. Despite the fact that there are two keyboards, one can only

Table 1.1 Input–output transactions table

		Consuming sectors			Final demand				Total output
		Agriculture	Industrial	Services	Consumption	Investments	Government expenditure	Net exports	
Producing sectors	Agriculture								
	Industrial								
	Services								
Value added	Compensation								
	Profit								
	Indirect taxes								
Total input									

assemble one typical computer station. In essence, the input–output model assumes that producing goods and services requires a fixed proportion of inputs. This leads to the assumption that the production function is specified as:

$$q = \min(x, y, z) \quad (1.1)$$

where q is the level of output, x is the amount of input x, y is the amount of input y, and z is the amount of input z. It should be noted that x, y, and z are perfect complements such that any excess resource will not contribute to an additional unit of output.

This assumption is restrictive in the sense that it does not allow sectors to substitute inputs for production. In our example, a typical computer station requires a definite specification; hence, the unavailability of one part cannot be substituted by a similar product. Furthermore, economic sectors do not benefit from economies of scale. These assumptions are justifiable in the short term under the condition of a fixed state of technology, which in turn restricts the flexibility of the actual physical input and output proportions for processes. In the long term, changes in capital stock can result in process technology changes that may result in new input–output coefficients or substitution. A typical component (i.e., economic sector) of any input–output system can be represented as a "black box" as shown in Fig. 1.2.

There is no supply constraint.

Increased level of final demand may result from policy reforms, trade liberalization, and changes in consumption preferences. These will generate an increase in intermediate demand in the affected economic sector thereby increasing the production requirement of other sectors without accounting for possible supply

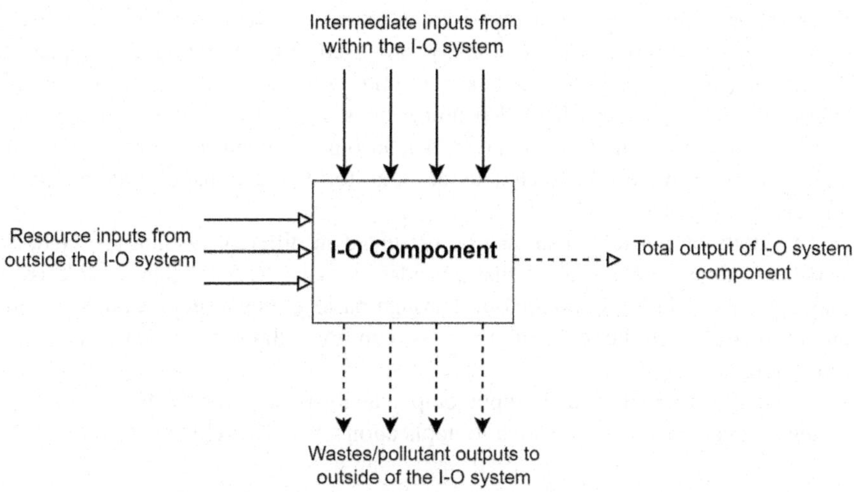

Fig. 1.2 Component of an input–output system

constraints. Resources are assumed to be available when the need arises. This condition in the basic model can be relaxed to account for limits (e.g., to the supply of natural resources) by adding additional equations or inequalities to reflect such constraints.

The model does not account for changes in price.

A basic economic concept requires price of goods and services to change alongside changes in demand or supply of the goods and services. Given that the input–output model seeks to estimate the impact of changes in final demand to total output, and an increase in final demand will bring about higher levels of total output, this should bring about a shift in the demand curve leading to higher equilibrium prices.

Although these assumptions are restrictive from the perspective of a thorough economic analysis, the simplifications ensure that the input–output model can be utilized as a means of tracking economic structures and making reasonable short-term predictions of economic systems. They also provide a more intuitive understanding of economic behavior, taking into account underlying technologies; the latter aspect is often missing in classical economic models. Such simplified models are essential to aid in the proper interpretation of complex phenomena described by detailed models [6].

1.3 Other Features of Input–Output Models

Scalability is one of the attractive features of input–output analysis, which is proven through the broad range of users of this tool. International agencies such as the Organisation for Economic Co-operation and Development (OECD), European Union (EU), and the Asian Development Bank (ADB) have published multi-regional input–output analysis to account for interdependencies between sectors across economies. These tables are used for estimating the impact of changes in policies and external shocks to trade flows across countries which will affect the economies involved depending on their dependence on imports and exports. Aside from input–output tables that account for countries as regions, input–output tables can be constructed to account for interactions of sectors among regions in a country.

The input–output table can also be presented in different resolutions. National statistical offices publish tables that consider as many as 500 sectors while some only consider as little as 200 sectors. Through basic aggregation techniques, input–output metrics can be compared across countries through standardizing their resolution.

While the original idea of input–output analysis was for studying economic systems, it can easily be extended to applications in other fields of study, such as:

- Economic–environmental systems—Interindustry interactions are captured through the traditional input–output model and a satellite matrix which accounts for the environmental emission intensity of each sector [8].
- Enterprise input–output models—The input–output framework is adapted to consider processes and resources required to manufacture output of a firm and measures interactions in physical terms [1].
- Organizational systems—Interactions between departments of organizations such as a library [3], a city municipal government [4], and a hospital [5] quantified through time are depicted through an input–output table.
- Food chains in ecosystems—The use of input–output analysis to elucidate the structure of ecosystems was first proposed by Hannon [7]. This application represents a radical departure from the conventional uses of input–output models to analyze man-made systems and underscores the versatility of this framework.

Indeed, the input–output analysis has proven its versatility not only through its application to the field of economics but also through its application to various fields of sciences. A subset of these variants of input–output models is covered in this book.

1.4 Overview of the Book

This book focuses on the use of input–output models for the design, planning, and analysis of sustainable industrial systems at different scales. The three prerequisite pillars of sustainability—economic viability, environmental soundness, and social acceptability—can all be potentially subsumed within a properly formulated input–output model, within the limiting assumptions discussed earlier. The related concept of system resilience, or the ability of systems to recover to normal state after being subjected to a shock, is also a recurring theme in this book. The most powerful feature of the input–output framework, its ability to reflect interdependencies among components in a network or system, is especially valuable due to the high level of complexity of modern industrial systems. Decisions made without the proper appreciation of indirect effects that can ripple through input–output systems can lead to adverse unexpected consequences, even if they are driven by the best of intentions.

The rest of this book is organized as follows. Chapter 2 discusses the mathematical foundations of input–output models, which is necessary for understanding the variants presented subsequently. Chapter 3 provides a brief overview of LINGO, which is the software platform used to solve all the examples in this book. Although it is possible to use spreadsheets for many of this model, LINGO has the advantage of using an equation-based interface and also features a versatile suite of solvers for optimization models.

Subsequent chapters discuss variants of the basic input–output model used for the analysis of economic structures. Chapter 4 covers input–output models of industrial complexes or clusters of firms linked in networks. Such models depart from traditional input–output models in the use of physical rather than monetary units to track flows of goods. Then, Chapter 5 discusses input–output models of infrastructure systems consisting of mutually interdependent components. Chapter 6 deals with input–output models of supply chain networks. This topic is of contemporary interest due to increasing concerns about the resilience of systems to climate change impacts. Input–output models provide the capability to project how economic systems are affected by disruptions and furthermore can be configured to give the best strategies for damage control. Life cycle assessment, a methodology for the analysis of the "cradle-to-grave" environmental impacts of product systems, uses a similar mathematical structure as input–output analysis. In a sense, any product system can be treated as a "slice" of a full economic system and can be analyzed as such. This topic is covered in Chapter 7. Chapter 8 covers the topic of input–output models for industrial plants. Subtopics such as early-stage design (i.e., synthesis), debottlenecking, and optimization of abnormal operating states are discussed. Chapter 9 discusses input–output models of organizational structures. This extension is based on the notion that workplace interactions within an organization are de facto economic transactions, measured in man-hours rather than explicit economic units. Organizational interactions among office units or employee categories can be seen as miniature economic systems and hence modeled as such. Finally, Chapter 10 discusses further research prospects on input–output models. Such prospects arise from the proven versatility of Leontief's framework, which should continue to find new applications in the analysis of linear or near-linear networks.

References

1. Albino V, Kuhtz S (2004) Enterprise input-output model for local sustainable development—the case of tiles manufacturer in Italy. Resour Conserv Recycl 41:165–176
2. Chen K (1973) Input-output economic analysis of environmental impact. IEEE Trans Syst Man Cybern 6:539–547
3. Correa H, Correa V (1996) An application of input-output analysis to the administration of a library. Libr Inf Sci Res 18:343–356
4. Correa H, Guajardo SA (2001) An application of input–output analysis to a city's municipal government. Socio-Econ Plann Sci 35:83–108
5. Correa H, Parker BR (2005) An application of organizational input–output analysis to hospital management. Socio-Econ Plann Sci 39:307–333
6. Geoffrion AM (1976) The purpose of mathematical programming is insight, not numbers. Interfaces 7:81–92
7. Hannon B (1973) The structure of ecosystems. J Theor Biol 41:535–546
8. Heijungs R, Suh S (2002) The computational structure of life cycle assessment. Springer, Dordrecht, Netherlands
9. Hubacek K, Giljum S (2003) Applying physical input-output analysis to estimate land appropriation (ecological footprints) of international trade activities. Ecol Econ 44:137–151
10. Leontief W (1936) Quantitative input and output relations in the economic system of the United States. Rev Econ Stat 18:105–125

Chapter 2
Mathematical Foundations of Input–Output Models

Abstract A detailed discussion of the mathematical foundations of input–output analysis is given in this chapter. The basic data format used in economic input–output models is discussed first, leading to the basic formulation and its solution using the matrix inversion approach. The use of the basic input–output model for key sector analysis is illustrated with the aid of an example. Applications of the input–output equations are depicted using spreadsheets, which provide an ideal foundation to understand the LINGO codes in subsequent chapters. Then, extensions involving mathematical programming, regional input–output models, and physical input–output models are considered; these variants are also illustrated with examples.

Keywords Leontief inverse · Key sector · Optimization · Mathematical programming · Regional input–output model · Physical input–output model

2.1 Introduction

The input–output (I–O) model views the economy as a set of interconnected sectors, which produce goods and consume goods in the process of production. Wassily Leontief won a Nobel Prize in Economics in 1973 for what became known as the Input–Output Model for Economy [10, 11]. Miller and Blair [12] provide a comprehensive introduction of the model and its applications. The model decomposes the output of an economic sector as a combination of intermediate consumptions and final demands. The model has been applied to a myriad of economic problems. It has proven to be a useful model for evaluating impacts of various structural changes across multiple sectors of national and regional economies [8]. The model is a useful tool in economic decision-making processes used in many countries because it is capable of describing the degree of interdependencies among various economic sectors and providing estimates of ripple effects associated with structural changes. Notably, extensions and new frontiers on input–output analysis can be found in [6]. The availability of high-resolution economic data and social

R. R. Tan et al., *Input–Output Models for Sustainable Industrial Systems*,
Lecture Notes in Management and Industrial Engineering,
https://doi.org/10.1007/978-981-13-1873-3_2

accounting matrices enables the application of the input–output model (and its extensions) for assessing the impact of consumption and production shifts in an economy. The objective of this chapter is to discuss the basic mathematical formulation of the input–output model and to give simple examples in order to better understand concepts such as technical coefficients and economic multipliers.

2.2 Formulation of the Input–Output Model

Wassily Leontief's input–output model provides useful information for studying the transactions among producers and consumers in the US national economy, focusing on the study of interconnected economic systems, where proportionality holds. The structure of the input–output model of the economy is summarized in Table 2.1.

Taking the sum of the last column:

$$X = x_1 + x_2 + L + N + M \tag{2.1}$$

Taking the sum of the last row:

$$X = x_1 + x_2 + C + I + G + E \tag{2.2}$$

Table 2.1 Basic Leontief I–O model structure [12]

		Processing sectors		Final demand				Total output
		1	2					
Processing sectors	1	z_{11}	z_{12}	C_1	I_1	G_1	E_1	x_1
	2	z_{21}	z_{22}	C_2	I_2	G_2	E_2	x_2
Payments	Value	L_1	L_2	L_C	L_I	L_G	L_E	L
Sector	Added	N_1	N_2	N_C	N_I	N_G	N_E	N
		M_1	M_2	M_C	M_I	M_G	M_E	M
Total input		x_1	x_2	C	I	G	E	X

where
z_{ij} is the input of sector i to j
C_i is the consumer (household) component of final demand for output of sector i
I_i is the investment (private) component of final demand for output of sector i
G_i is the government component of final demand for output of sector i
E_i is the export component of final demand for output of sector i
x_i is the total output of sector i
L_j is the labor input for sector j
N_j is the other value-added input for sector j
M_j is the import component of input for sector j
X is the total output for the entire economy

Combining Eqs. (2.1) and (2.2):

$$L + N + M = C + I + G + E \tag{2.3}$$

Rearranging Eq. (2.3) will result in the following fundamental economic identity:

$$L + N = C + I + G + (E - M) \tag{2.4}$$

The left-hand side of Eq. (2.4) represents the gross national income, while the right-hand side represents the gross national product.

In order to derive the basic input–output equations, suppose that an economy consists of n interacting sectors. We can extract relevant information from Table 2.1 and generalize the number of processing sectors to n. This will result in the following variables and parameters for the Leontief economic input–output model formulation.

- z_{ij}: input of industry i to industry j (intermediate consumption)
- a_{ij}: input of industry i to j, normalized with respect to the total output of industry j
- f_i: final demand for industry i
- x_i: total output of industry i
- x_j: total output of industry j.

where $i, j = 1, 2, \ldots, n$

The proportionality assumption leads to the following equation.

$$z_{ij} = a_{ij} x_j \tag{2.5}$$

Furthermore, the following balance equation suggests that the total output of industry i is consumed either as intermediate demands (i.e., z_{ij}) or as final demand (f_i). For example, suppose that industry i produces sugar. The output of industry i (i.e., sugar industry) can either be used as an intermediate input to industry j (e.g., confectionery industry) or consumed as a final product (e.g., sugar consumed in households). Such allocation of an industry's output to various consumers (intermediate and final) translates to the following mathematical formulation.

$$x_i = \sum_{j=1}^{n} z_{ij} + f_i \tag{2.6}$$

Substituting Eq. (2.5) into Eq. (2.6) will reveal the basic Leontief input–output model.

$$x_i = \sum_{j=1}^{n} a_{ij} x_j + f_i \tag{2.7}$$

In matrix form, Eq. (2.3) can be written as follows.

$$\mathbf{x} = \mathbf{A}\mathbf{x} + \mathbf{f} \tag{2.8}$$

In the matrix notation of Eq. (2.8), the variables are interpreted similarly as their scalar counterparts: **x** is the total output vector (a column vector representing the total output of each industry), **f** is the final demand vector (a column vector representing the final demand for each industry), and **A** is a square matrix whose elements represent the proportion of the input of industry i to j with respect to the total output of industry j. In input–output literature, **A** is typically called the Leontief technical coefficient matrix.

In the following equations, we will show how to derive and describe the interpretations for the Leontief inverse, typically denoted in the literature by **L**. Using Eq. (2.4) as the starting point, the aim is to explicitly isolate **x** on the left side of the equation. We do this through the following steps.

$$\mathbf{x} - \mathbf{A}\mathbf{x} = \mathbf{f} \tag{2.9}$$

$$(\mathbf{I} - \mathbf{A})\mathbf{x} = \mathbf{f} \tag{2.10}$$

$$\mathbf{x} = (\mathbf{I} - \mathbf{A})^{-1}\mathbf{f} \tag{2.11}$$

Note that **I** is an identity matrix with the same size as **A**. In Eq. (2.11), we can define **L** as the inverse term:

$$\mathbf{L} = (\mathbf{I} - \mathbf{A})^{-1} \tag{2.12}$$

Substituting Eq. (2.12) into Eq. (2.11) will reveal an even more simplified version of the input–output model:

$$\mathbf{x} = \mathbf{L}\mathbf{f} \tag{2.13}$$

Note that the inverse term $(\mathbf{I} - \mathbf{A})^{-1}$, which is denoted by **L**, is often referred to in the literature as the Leontief inverse. It is also called the total requirements matrix, which will be revisited further in this chapter.

Example 2.1 Calculate the **A** matrix and the **L** Matrix given the following hypothetical two-sector economy as shown below and illustrated in Fig. 2.1 (Table 2.2).

Solution
The total output (**x**) and final demand (**f**) vectors can be extracted directly from the table.

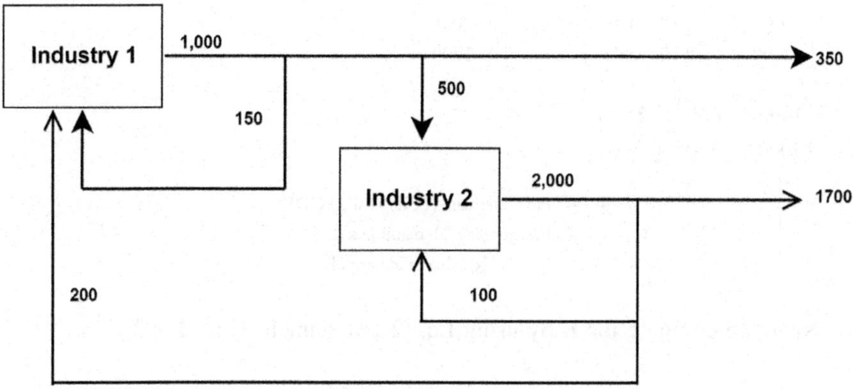

Fig. 2.1 Two-sector economy

Table 2.2 Hypothetical two-sector economy (adapted from [12])

Industries	1	2	Final demand	Total output
1	**150**	**500**	350	1000
2	**200**	**100**	1700	2000
Value added	650	1400		
Total input	1000	2000		

$$\mathbf{x} = \begin{bmatrix} 1000 \\ 2000 \end{bmatrix} \quad \mathbf{f} = \begin{bmatrix} 350 \\ 1700 \end{bmatrix}$$

The Leontief technical coefficient matrix (**A**) is determined by normalizing the shaded cells by the corresponding column sum:

$$\mathbf{A} = \begin{bmatrix} 150/1000 & 500/2000 \\ 200/1000 & 100/2000 \end{bmatrix} = \begin{bmatrix} 0.15 & 0.25 \\ 0.20 & 0.05 \end{bmatrix}$$

Using Eq. (2.8), we shall be able to verify that indeed both sides of the equation match.

$$\begin{bmatrix} 1000 \\ 2000 \end{bmatrix} = \begin{bmatrix} 0.15 & 0.25 \\ 0.20 & 0.05 \end{bmatrix} \begin{bmatrix} 1000 \\ 2000 \end{bmatrix} + \begin{bmatrix} 350 \\ 1700 \end{bmatrix}$$

$$\begin{bmatrix} 1000 \\ 2000 \end{bmatrix} = \begin{bmatrix} 0.15 & 0.25 \\ 0.20 & 0.05 \end{bmatrix} \begin{bmatrix} 1000 \\ 2000 \end{bmatrix} + \begin{bmatrix} 350 \\ 1700 \end{bmatrix}$$

$$\begin{bmatrix} 1000 \\ 2000 \end{bmatrix} = \begin{bmatrix} 650 \\ 300 \end{bmatrix} + \begin{bmatrix} 350 \\ 1700 \end{bmatrix}$$

- Final Demand (**f**), also called GDP
- Intermediate Demand (**Ax**)
- Output (**x**), also called Supply

Next, we compute for **L** by using Eq. (2.12), namely $\mathbf{L} = (\mathbf{I} - \mathbf{A})^{-1}$:

$$\mathbf{L} = (\mathbf{I} - \mathbf{A})^{-1} = \left(\begin{bmatrix} 1 & 0 \\ 0 & 1 \end{bmatrix} - \begin{bmatrix} 0.15 & 0.25 \\ 0.20 & 0.05 \end{bmatrix} \right)^{-1} = \begin{bmatrix} 0.85 & -0.25 \\ -0.20 & 0.95 \end{bmatrix}^{-1}$$

The inverse can be computed using standard mathematical software packages, such as LINGO, which will be explored in subsequent chapters of this book. Note that MS Excel spreadsheet is also capable of matrix inversion via the "MINVERSE" function. The resulting Leontief inverse can be computed as follows:

$$\mathbf{L}{=}(\mathbf{I} - \mathbf{A})^{-1} = \begin{bmatrix} 0.85 & -0.25 \\ -0.20 & 0.95 \end{bmatrix}^{-1}$$

$$\mathbf{L} = \begin{bmatrix} 1.25 & 0.33 \\ 0.26 & 1.12 \end{bmatrix}$$

Matrix **L** is a relatively straightforward concept, yet it is considered one of the most powerful strengths of the I-–O model. To interpret the elements of the **L** matrix, suppose that we are using $ as the currency unit (noting that the following interpretations would also work for other types of currencies).

$$\mathbf{L} = \begin{bmatrix} 1.25 & 0.33 \\ 0.26 & 1.12 \end{bmatrix}$$

For every $1 increase in the Demand for Industry 1:
- Output for Industry 1 will increase by $1.25
- Output for Industry 2 will increase by $0.26

Total Output for both industries will increase by $1.51

$$\mathbf{L} = \begin{bmatrix} 1.25 & 0.33 \\ 0.26 & 1.12 \end{bmatrix}$$

For every $1 increase in the Demand for Industry 2:
- Output for Industry 1 will increase by $0.33
- Output for Industry 2 will increase by $1.12

Total Output for both industries will increase by $1.45

Now, suppose that the government were to stimulate an increase in the demand for a sector, which sector gives the higher impact in terms of total output increase? This question can be answered by taking the column sums of the **L** matrix, which gives the total output multipliers.

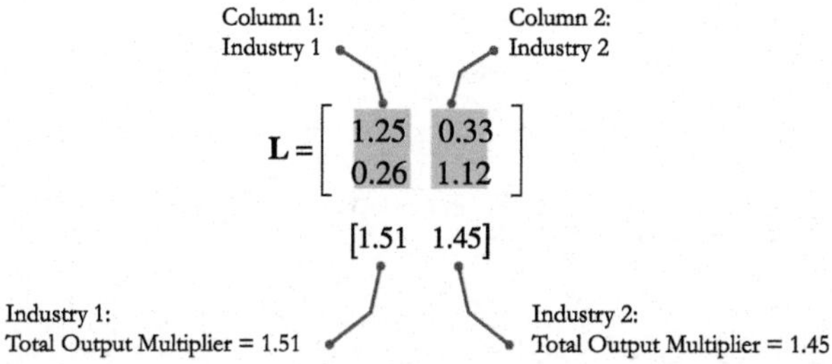

Total output multipliers provide the most commonly used metric for identifying the key sectors, which will be particularly useful in the analysis of higher-order matrices (i.e., more industry sectors).

2.3 Key Sector Analysis

One important result of a study of interdependent systems is the ability to have a better understanding of the system components (e.g., economic sectors) that are most interconnected with other components. A measure of the interconnectedness of an economic sector may be of interest to a policymaker who wishes to maximize the effect of a structural shift to that sector. For example, the amount of usage and interconnectedness of certain transportation components may provide policy insights into making investment priorities. A study of the highly connected components of a power grid can provide an understanding of why blackouts, such as the occurrence in the northeast USA in August 2003, can occur with widespread consequence [1]. In general, a country interested in increasing its growth rate studying highly interconnected sectors that have the ability to spread growth along the entire economy will benefit from this evaluation.

Hence, one of the major topics within the domain of I–O modeling is the identification, estimation, and prioritization of the key sectors in an economy. The term "key sector" describes an industry with strong influence on the expansion of others in an economy [9], and several metrics have been proposed to identify these key sectors. Key sectors have been identified based on column sums in the technical coefficients matrix **A** [4] and in the Leontief inverse **L** [13]. Cella [3] discussed the hypothetical extraction method, where the difference between a standard n-sector Leontief inverse and some reduced Leontief inverse is studied. Dietzenbacher [5] shows that the elements of eigenvector corresponding to the dominant eigenvalue of the technical coefficient matrix can provide a method for ordering key sectors. The multiplier product matrix [14] has been proposed to generate a hierarchy of technical coefficients based on the first-order field of influence, which identifies the important elements of a matrix.

Example 2.2 You are given the 2016 total requirements matrix for the USA below. Note that this is also the same as the Leontief inverse (**L**). Which sector gives the highest and lowest total output multipliers? Analyze these total multiplier values using the concept of key sector analysis (Table 2.3).

Table 2.3 Total requirements multipliers for 2016 US economy

Industry	Agriculture	Mining	Utilities	Construction	Manufacturing	Wholesale	Retail	Transportation
Agriculture	1.30	0.01	0.00	0.02	0.09	0.00	0.01	0.01
Mining	0.03	1.09	0.09	0.03	0.08	0.00	0.01	0.01
Utilities	0.02	0.01	1.01	0.01	0.01	0.00	0.01	0.01
Construction	0.01	0.02	0.02	1.00	0.01	0.00	0.01	0.01
Manufacturing	0.35	0.14	0.09	0.36	1.57	0.06	0.08	0.22
Wholesale	0.10	0.03	0.02	0.07	0.09	1.03	0.03	0.05
Retail	0.01	0.00	0.00	0.07	0.01	0.00	1.01	0.02
Transportation	0.07	0.04	0.06	0.04	0.06	0.05	0.06	1.15
Information	0.02	0.01	0.01	0.02	0.03	0.03	0.04	0.03
Finance	0.13	0.06	0.05	0.07	0.07	0.10	0.16	0.14
Professional	0.09	0.10	0.08	0.11	0.18	0.16	0.17	0.14
Educational	0.00	0.00	0.00	0.00	0.00	0.00	0.01	0.00
Entertainment	0.01	0.01	0.01	0.01	0.01	0.01	0.01	0.01
Other services	0.01	0.00	0.00	0.01	0.01	0.01	0.01	0.01
Government	0.02	0.01	0.01	0.01	0.02	0.02	0.02	0.05
Totals	2.16	1.51	1.46	1.83	2.26	1.51	1.63	1.86

Industry	Information	Finance	Professional	Educational	Entertainment	Other services	Government
Agriculture	0.01	0.00	0.01	0.01	0.02	0.01	0.01
Mining	0.01	0.01	0.01	0.01	0.01	0.01	0.02
Utilities	0.01	0.01	0.01	0.01	0.01	0.01	0.01
Construction	0.01	0.03	0.00	0.01	0.01	0.01	0.03
Manufacturing	0.17	0.05	0.10	0.15	0.22	0.16	0.18
Wholesale	0.04	0.01	0.02	0.03	0.04	0.03	0.03
Retail	0.00	0.00	0.00	0.00	0.01	0.02	0.00

(continued)

Table 2.3 (continued)

Industry	Information	Finance	Professional	Educational	Entertainment	Other services	Government
Transportation	0.03	0.01	0.03	0.02	0.03	0.02	0.03
Information	1.21	0.03	0.05	0.04	0.04	0.03	0.05
Finance	0.10	1.22	0.13	0.19	0.14	0.17	0.07
Professional	0.17	0.13	1.22	0.17	0.19	0.12	0.14
Educational	0.00	0.00	0.00	1.01	0.00	0.00	0.01
Entertainment	0.04	0.02	0.03	0.02	1.03	0.01	0.02
Other services	0.01	0.01	0.01	0.02	0.01	1.01	0.01
Government	0.02	0.02	0.01	0.02	0.02	0.02	1.02
Totals	1.82	1.56	1.63	1.72	1.79	1.63	1.61

Source [2]

Table 2.4 Ranking of key sectors based on total output multipliers

Rank	Sector	Total output multiplier
1	Manufacturing	2.255
2	Agriculture	2.163
3	Transportation	1.857
4	Construction	1.830
5	Information	1.815
6	Entertainment	1.787
7	Educational	1.721
8	Other services	1.633
9	Professional	1.629
10	Retail	1.625
11	Government	1.610
12	Finance	1.555
13	Mining	1.510
14	Wholesale	1.507
15	Utilities	1.458

Solution

The column sums of the above table give the total output multipliers for every $1 increase in the demand for each column sector. The highest value is generated by the manufacturing sector, that is, for every $1 increase in the demand for manufacturing sector; the total output for the entire economy will increase by $2.26. Using the total output multiplier values as the basis for key sector analysis (i.e., the last row of the table), then the manufacturing sector is deemed to have the highest rank.

By the same logic, the lowest value is generated by the utility sector. To wit, every $1 increase in the demand for utilities, the total output for the entire economy will increase by $1.46. Hence, utility sector will have the lowest rank in the key sector analysis on the basis of the total output multiplier values. The complete list of key sector rankings based on the magnitude of total output multipliers is presented in Table 2.4.

2.4 Input–Output Model with Optimization

Recall that the basic Leontief I–O model follows the form:

$$\mathbf{x} = \mathbf{A}\mathbf{x} + \mathbf{f} \tag{2.14}$$

where

x is the production output vector
A is the matrix of technical coefficients
f is the final demand vector

From Eq. (2.14), we can isolate **f** on the left side of the equation to produce:

$$\mathbf{f} = (\mathbf{I} - \mathbf{A})\mathbf{x} \tag{2.15}$$

Defining **p** as the price vector of the final demand **f**, then the gross domestic product (GDP) is defined as the product of the two, namely:

$$\mathrm{GDP} = \mathbf{pf} \tag{2.16}$$

Hence,

$$\mathrm{GDP} = \mathbf{p}(\mathbf{I} - \mathbf{A})\mathbf{x} \tag{2.17}$$

For an economy where some resources are constrained, defining **B** as the resource coefficient matrix and **r** as the vector of upper limits for the available resources [7] will yield the following set of constraints.

$$\mathbf{Bx} \leq \mathbf{r} \tag{2.18}$$

2.4.1 I–O Model with Resource Constraints: Primal Problem

Combining Eqs. (2.17) and (2.18) and imposing a nonnegativity constraint for the production output **x**, we can form an optimization problem where we maximize the GDP:

$$\begin{aligned} &\text{Max} \quad \mathrm{GDP} = \mathbf{p}(\mathbf{I} - \mathbf{A})\mathrm{x} \\ &\text{subject to:} \\ &\mathbf{Bx} \leq \mathbf{r} \\ &\mathbf{x} \geq \mathbf{0} \end{aligned} \tag{2.19}$$

2.4.2 I–O Model with Resource Constraints: Dual Problem

Let **w** be the cost vector associated with the resource constraints. The dual problem seeks to minimize the total cost which is the product of **w** and **r**. The resulting dual formulation of Eq. (2.19) will be:

$$\begin{aligned}&\text{Min}\quad \mathbf{wr}\\&\text{subject to:}\\&\mathbf{wB} \geq \mathbf{p}(\mathbf{I}-\mathbf{A})\\&\mathbf{w} \geq \mathbf{0}\end{aligned} \tag{2.20}$$

2.4.3 Summary: Dual and Primal Optimization Problems

Define an adjusted price vector:

$$\bar{\mathbf{p}} = \mathbf{p}(\mathbf{I}-\mathbf{A}) \tag{2.21}$$

The optimization problems for the primal and dual formulations become:

Primal formulation	Primal formulation
Max $\bar{\mathbf{p}}\mathrm{x}$	Min $\mathbf{wr}$
subject to:	subject to :
$\mathbf{Bx} \leq \mathbf{r}$	$\mathbf{wB} \geq \bar{\mathbf{p}}$
$\mathbf{x} \geq \mathbf{0}$	$\mathbf{w} \geq \mathbf{0}$

Getting the Lagrangian of the primal problem:

$$L(\mathbf{x};\mathbf{w}) = -\bar{\mathbf{p}}\mathbf{x} + \mathbf{w}(\mathbf{Bx}-\mathbf{r}) \tag{2.22}$$

Applying a necessary Kuhn–Tucker condition for optimality, where **x*** is the optimal solution for the primal problem (assuming it exists):

$$\mathbf{w}^{*}\frac{\partial L(\mathbf{x}^{*};\mathbf{w}^{*})}{\partial \mathbf{w}} = \mathbf{w}^{*}(\mathbf{Bx}^{*}-\mathbf{r}) = 0 \tag{2.23}$$

Getting the Lagrangian of the dual problem:

$$L(\mathbf{w};\mathbf{x}) = \mathbf{wr} - (\mathbf{wB}-\bar{\mathbf{p}})\mathbf{x} \tag{2.24}$$

Applying a necessary Kuhn–Tucker condition for optimality, where **w*** is the optimal solution for the dual problem (assuming it exists):

$$\mathbf{x}^{*}\frac{\partial L(\mathbf{w}^{*};\mathbf{x}^{*})}{\partial \mathbf{x}} = -(\mathbf{w}^{*}\mathbf{B}-\bar{\mathbf{p}})\mathbf{x}^{*} = 0 \tag{2.25}$$

Combining Eqs. (2.23) and (2.25):

$$\mathbf{w}^*(\mathbf{B}\mathbf{x}^* - \mathbf{r}) - (\mathbf{w}^*\mathbf{B} - \bar{\mathbf{p}})\mathbf{x}^* = 0 \tag{2.26}$$

Simplifying Eq. (2.26):

$$\bar{\mathbf{p}}\mathbf{x}^* = \mathbf{w}\mathbf{r}^* \tag{2.27}$$

Hence, we have shown that the optimal values of the objective functions for the primal and dual problems must equal each other (assuming no duality gap).

Example 2.3
Revisit the two-sector technical coefficient matrix that we previously considered in Example 2.1.

$$\mathbf{A} = \begin{bmatrix} 0.15 & 0.25 \\ 0.2 & 0.05 \end{bmatrix}$$

Assume there are two resources whose input structure is assumed fixed:

$$\mathbf{B} = \begin{bmatrix} 0.4 & 0.1 \\ 0.1 & 0.4 \end{bmatrix}$$

Furthermore, suppose that the maximum available quantities of resources 1 and 2 are as follows:

$$\mathbf{r} = \begin{bmatrix} 600 \\ 900 \end{bmatrix}$$

Solve for the optimal production output of sectors 1 and 2 by maximizing the GDP. Assume that the price vector of the corresponding outputs is $\mathbf{p} = [1 \quad 1]$

Solution
Given:

$$\mathbf{A} = \begin{bmatrix} 0.15 & 0.25 \\ 0.2 & 0.05 \end{bmatrix}$$

$$\mathbf{p} = [1 \quad 1]$$

Get $\mathbf{I} - \mathbf{A}$

$$\mathbf{I} - \mathbf{A} = \begin{bmatrix} 0.85 & -0.25 \\ -0.2 & 0.95 \end{bmatrix}$$

Get **p(I − A)x**

$$\mathbf{p}(\mathbf{I}-\mathbf{A})\mathbf{x} = \begin{bmatrix} 1 & 1 \end{bmatrix} \begin{bmatrix} 0.85 & -0.25 \\ -0.2 & 0.95 \end{bmatrix} \begin{bmatrix} x_1 \\ x_2 \end{bmatrix}$$
$$\mathbf{p}(\mathbf{I}-\mathbf{A})\mathbf{x} = 0.65x_1 + 0.7x_2$$

Primal Optimization Problem:

$$\text{Max} \quad \begin{bmatrix} 0.65 & 0.7 \end{bmatrix} \begin{bmatrix} x_1 \\ x_2 \end{bmatrix}$$
$$\text{subject to:}$$
$$\begin{bmatrix} 0.4 & 0.1 \\ 0.1 & 0.4 \end{bmatrix} \leq \begin{bmatrix} 600 \\ 900 \end{bmatrix}$$
$$\begin{bmatrix} x_1 \\ x_2 \end{bmatrix} \geq \begin{bmatrix} 0 \\ 0 \end{bmatrix}$$

Optimal Solution:

$$\text{GDP} = 2050$$
$$\begin{bmatrix} x_1 \\ x_2 \end{bmatrix} = \begin{bmatrix} 1000 \\ 2000 \end{bmatrix}$$

Analysis: The optimal production for sectors 1 and 2 are 1000 and 2000, respectively, for a total GDP of 2050. Since no unit was explicitly specified, we may assume that the GDP value of 2050 is in monetary units.

Exercise
Use the dual formulation and obtain the optimal values of the shadow prices for resources 1 and 2.

2.5 Regional Input–Output Model

Regional decomposition enables a more focused and thus more accurate analysis of infrastructure interdependencies for desired regions in the USA. Miller and Blair [12] discuss the case of “closing” the I–O analysis to a particular region (i.e., a single regional I–O framework as opposed to multiregional). At the national level, the I–O model is **x = Ax + f**. The regional model for region R takes a similar form.

$$\mathbf{x}^R = \mathbf{A}^R\mathbf{x}^R + \mathbf{f}^R \tag{2.28}$$

The system of equations corresponding to Eq. (2.28) is as follows:

$$\begin{aligned} x_1^R &= a_{11}^R x_1^R + a_{12}^R x_2^R + \cdots + a_{1n}^R x_n^R + f_1^R \\ x_2^R &= a_{21}^R x_1^R + a_{22}^R x_2^R + \cdots + a_{2n}^R x_n^R + f_2^R \\ &\vdots \\ x_n^R &= a_{n1}^R x_1^R + a_{n2}^R x_2^R + \cdots + a_{nn}^R x_n^R + f_n^R \end{aligned} \tag{2.29}$$

2.5.1 Location Quotients

With the availability of national I–O tables, analysts can convert and customize them to produce regional I–O tables using the economic characteristics of the region (e.g., sector productions, incomes, regional purchases). For example, [8] and Miller and Blair [12] define a simple location quotient (l_i) using the following ratio:

$$l_i = \frac{x_i^R / x_s^R}{x_i / x_s} \tag{2.30}$$

where

x_i^R is the regional output for the ith industry
x_s^R is the total regional output for all regional-level industries
x_i is the national output for the ith industry
x_s is the total national output for all national-level industries

The regional industry-by-industry technical coefficient matrix $\mathbf{A}^R$, whose elements are denoted by a_{ij}^R, is then established as follows.

$$a_{ij}^R = \begin{cases} a_{ij}(l_i) & l_i < 1 \\ a_{ij} & l_i \geq 1 \end{cases} \tag{2.31}$$

When ℓ is used to denote a vector of location quotients and $\mathbf{i}$ a unity vector, Eq. (2.31) can be written in the following matrix notation.

$$\mathbf{A}^R = \operatorname{diag}(\operatorname{Min}(\ell, \mathbf{i}))\mathbf{A} \tag{2.32}$$

In scalar form, the above equation translates to:

$$a_{ij}^R = \operatorname{Min}(\ell_i, 1)a_{ij} \tag{2.33}$$

2.6 Physical Input–Output Model

A vast majority of I–O matrices are measured in terms of monetary units. Nonetheless, there are special cases where the data are provided in their "natural" (or physical) units. For example, we have observed from the earlier section (I–O model with optimization) the possibility of embedding resource constraints that is expressed in terms of their natural units. In this section, we will explore the case where the technical coefficient matrix (**A**) originates from a matrix of commodities expressed in their natural units. Consider the physical I–O model equivalent of the earlier hypothetical two-sector economy (Table 2.5).

It should be noted that, unlike the I–O data with monetary units, it is invalid to take the column sums of the above table because the units are different. Nonetheless, taking the sum across each row is valid since the units are consistent (e.g., the first row is in bushels, the second row is in tons). Now, suppose that we know the price per bushel of commodity 1, the price per ton of commodity 2, and the price per hour of the value added. The unit prices are shown in Table 2.6.

Multiplying the unit price data to the corresponding rows of the commodity flow table shown previously will result in a monetary I–O data, where all cells are now expressed in dollar units (Table 2.7).

Simplifying the above table will reveal its equivalence to the monetary I–O data shown earlier in this chapter. Note that since all units are in dollars, it would now be valid to take the column sums since the natural units of the commodities and value added already canceled out when the products with unit prices were performed (Table 2.8).

The above derivations show the fundamental equivalence in monetary I–O table and physical I–O table, with unit price serving as the link between the two tables. A generalization can be made to show such equivalence.

Let

- y_i: total commodity output of industry i (in natural units)
- y_j: total commodity output of industry j (in natural units)

Table 2.5 Flow of commodities in a hypothetical two-sector economy (adapted from [12])

Commodities	1	2	Final demand	Total output	Natural units
1	**75**	**250**	175	500	Bushels
2	**40**	**20**	340	400	Tons
Value added	65	140			Hours

Table 2.6 Unit price data

Commodities	Natural units	Unit price
1	Bushels	\$2/bushel
2	Tons	\$5/ton
Value added	Hours	\$10/h

Table 2.7 Flow of commodities in a hypothetical two-sector economy

Commodities	1	2	Final demand	Total output
1	**75 bushels × $2/bushel**	**250 bushels × $2/bushel**	175 bushels × $2/bushel	500 bushels × $2/bushel
2	**40 tons × $5/ton**	**20 tons × $5/ton**	340 tons × $5/ton	400 tons × $5/ton
Value added	65 h × $10/h	140 h × $10/h		

Table 2.8 Flow of commodities in a hypothetical two-sector economy, expressed in monetary units

Industries	1	2	Final demand	Total output
1	**$150**	**$500**	$350	$1000
2	**$200**	**$100**	$1700	$2000
Value added	$650	$1400		
Total input	$1000	$2000		

- d_i: final commodity demand for industry i (in natural units)
- p_i: unit price of commodity i
- n_{ij}: commodity input of industry i to industry j (intermediate consumption, in natural units)
- c_{ij}: commodity input of industry i to j, normalized with respect to the total commodity output of industry i.

where $i, j = 1, 2, \ldots, n$.

Hence, the following balance equation suggests that the total commodity output of industry i is consumed either as intermediate commodity demands (i.e., n_{ij}) or as final commodity demand (d_i). Such allocation of an industry's commodity output to various consumers (intermediate and final) translates to the following mathematical formulation.

$$y_i = \sum_{j=1}^{n} n_{ij} + d_i \quad (2.34)$$

We assume that the proportionality assumption also works for the physical I–O data, which would result in the following ratio:

$$c_{ij} = n_{ij}/y_j \quad (2.35)$$

Substituting the above ratio into the previous equation will reveal the following equation.

$$y_i = \sum_{j=1}^{n} c_{ij}y_j + d_i \tag{2.36}$$

Note that in matrix form, the balance equation for the physical I–O model is structurally similar to its monetary counterpart.

$$\mathbf{y} = \mathbf{Cy} + \mathbf{d} \tag{2.37}$$

There is also a corresponding inverse, which can be obtained by isolating y on the left side of the above equation.

$$\mathbf{y} = (\mathbf{I} - \mathbf{C})^{-1}\mathbf{d} \tag{2.38}$$

In order to relate the physical I–O with the monetary I–O, we can multiply each term of Eq. (2.34) with p_i.

$$p_i y_i = \sum_{j=1}^{n} p_i n_{ij} + p_i d_i \tag{2.39}$$

The above equation reveals the relationship between the physical and monetary I–O models.

$$x_i = p_i y_i \tag{2.40}$$

$$z_{ij} = p_i n_{ij} \tag{2.41}$$

$$f_i = p_i d_i \tag{2.42}$$

Another interesting derivation is achieved by taking the column sums of the price-adjusted physical I–O, which would include the value-added row.

$$p_j y_j = \sum_{i=1}^{n} p_i n_{ij} + p_{n+1} n_{n+1,j} \tag{2.43}$$

where p_{n+1} is the unit price of value added, and $n_{n+1,j}$ corresponds to the last row of Table 2.5 (i.e., value added, which in this example was specifically allocated to labor for simplicity). Now, let us divide each term of the above equation by y_j.

$$p_j \frac{y_j}{y_j} = \sum_{i=1}^{n} p_i \frac{n_{ij}}{y_j} + p_{n+1} n_{n+1,j}/y_j \tag{2.44}$$

Note that $c_{ij} = n_{ij}/y_j$ and for simplicity, define $v_j = p_{n+1} n_{n+1,j}/y_j$. The above equation then becomes:

$$p_j = \sum_{i=1}^{n} p_i c_{ij} + v_j \tag{2.45}$$

In matrix form, the price I–O model, which was derived from the physical I–O model, can be written in matrix form as follows:

$$\mathbf{p}' = \mathbf{p}'\mathbf{C} + \mathbf{v}' \tag{2.46}$$

Isolating the price variable on the left side of the equation will yield:

$$\mathbf{p}' = \mathbf{v}'(\mathbf{I} - \mathbf{C})^{-1} \tag{2.47}$$

Analogous to the Leontief inverse matrix (**L**), the new inverse $(\mathbf{I} - \mathbf{C})^{-1}$ gives the price multiplier effect for every \$1 change in the price of value added.

2.7 Conclusion

In summary, this chapter introduces several fundamental concepts that will serve as the foundation for subsequent chapters of this book. Variants of the basic formulation are also discussed. Other related extensions of input–output models are discussed in the chapters that follow.

References

1. Anderson CW, Santos JR, Haimes YY (2007) A risk-based input-output methodology for measuring the effects of the August 2003 Northeast Blackout. Econ Syst Res 19(2):183–204
2. Bureau of Economic Analysis (2016) Input-output accounts data. Available at: https://www.bea.gov/industry/io_annual.htm. Accessed 4 Mar 2018
3. Cella G (1984) The input-output measurement of interindustry links. Oxford Bull Econ Stat 46(1):73–84
4. Chenery HB, Watanabe T (1958) International comparisons of the structure of production. Econometrica 26(4):98–139
5. Dietzenbacher (1992) The measurement of interindustry linkages: key sectors in the Netherlands. Econ Model 9(4):419–437
6. Dietzenbacher E, Lahr ML (2004) Wassily Leontief and input-output economics. Cambridge University Press, Cambridge, UK
7. Intriligator MD (1971) Mathematical optimization and economic theory. Prentice-Hall Inc, Englewood Cliffs, NJ
8. Isard W (1960) Methods of regional analysis: an introduction to regional science. MIT Press, Cambridge, MA
9. Lahr ML, Dietzenbacher E (2001) Input-output analysis: frontiers and extensions. Palgrave, New York, NY

10. Leontief WW (1951) The structure of the American Economy, 1919–1939, 2nd edn. Oxford University Press, New York, NY
11. Leontief WW (1966) Input-output economics. Oxford University Press, New York, NY
12. Miller RE, Blair PD. (2009) Input-output analysis: foundations and extensions. Cambridge University Press
13. Rasmussen P (1956) Studies in intersectoral relations. North Holland, Amsterdam, The Netherlands
14. Sonis M, Hewings GJD (1998) Temporal Leontief inverse. Macroecon Dyn 2(1):89–114

Chapter 3
Programming in LINGO

Abstract A brief introduction to the commercial optimization software LINGO is given. General features of the software are discussed, including user interface, file formats, and syntax. A generic example of an optimization model coded in LINGO is first given, followed by a specific case study that implements the basic input–output model. The use of set-based coding for large models is also introduced.

Keywords Optimization software · Set-based model

3.1 Introduction

LINGO is a specialized optimization software developed by LINDO Systems. It allows its user to build optimization models using an equation-based interface. Different classes of mathematical programming models can also be solved using different solver features. A free demo version, with limited capabilities, is available for download from https://www.lindo.com/, while a serious programmer can opt to purchase a license for the basic software and additional toolboxes to handle special classes of models. More details may be found in the Web site of LINDO Systems [1] and in Schrage [2]. The modeling language of LINGO is straightforward which makes it easy to build and express various mathematical models. This chapter provides an introduction to the basics of modeling in LINGO. Examples of more complex models and software features can be found at https://www.lindo.com/downloads/PDF/LINGO.pdf.

3.2 Installing the Software

LINGO software can be downloaded from LINDO Systems [1] in the downloads section of the Web site as shown in Fig. 3.1. The user will then be requested to select the platform version (Fig. 3.2). Once the software has been downloaded,

R. R. Tan et al., *Input–Output Models for Sustainable Industrial Systems*,
Lecture Notes in Management and Industrial Engineering,
https://doi.org/10.1007/978-981-13-1873-3_3

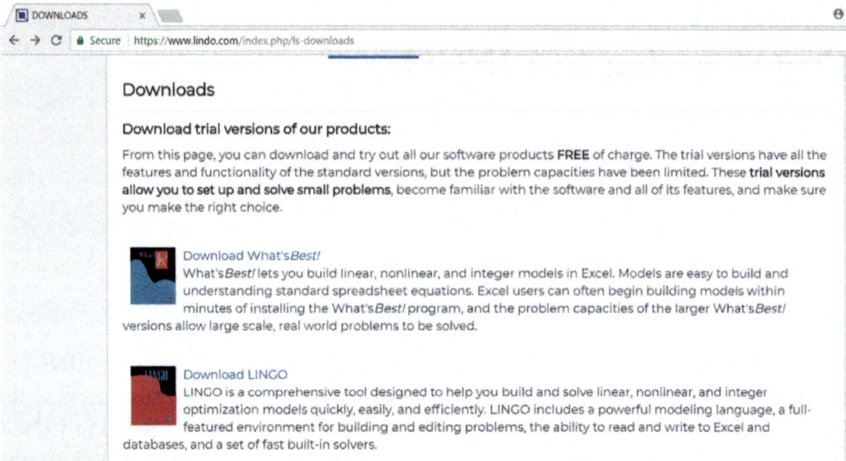

Fig. 3.1 Screenshot of the downloads section in the LINDO Systems' Web site

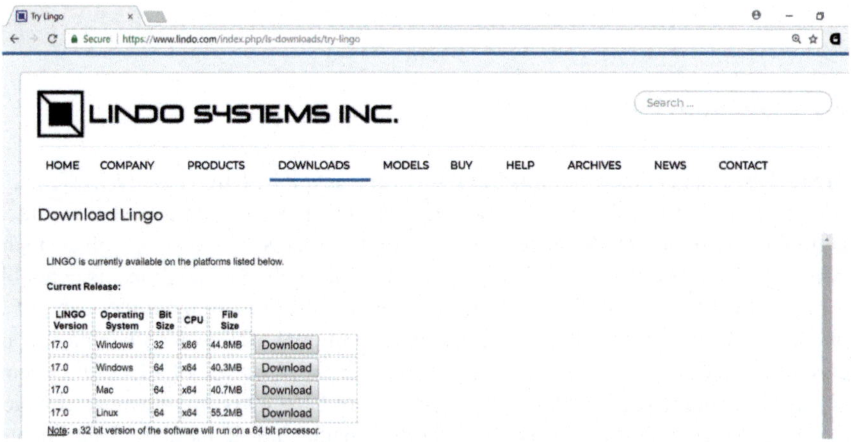

Fig. 3.2 Downloading the appropriate LINGO platform

LINGO can be installed by running the installation program SETUP. An icon will appear once the installation is successful. To launch the program, just double-click the icon and this should open a window as shown in Fig. 3.3. Figure 3.3 shows the main display window of LINGO, and this is where the optimization models are built.

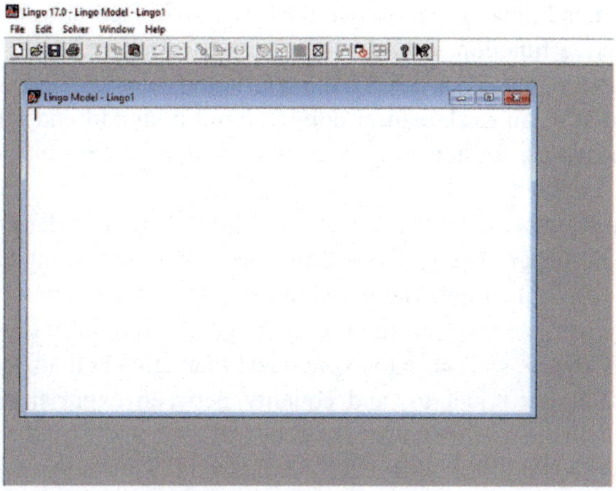

Fig. 3.3 Screenshot of the main display window of LINGO

3.3 Modeling in LINGO

Modeling in LINGO is straightforward, such that model components (e.g., equations, inequalities) are encoded in the same way that they are written in normal text. LINGO is not case sensitive and will not differentiate between the variable names VAR, var, vAr, Var, etc. There is no need to declare the variables prior to their use except when the set-based formulation is used (to be discussed in the later part of this section). Furthermore, LINGO assumes that all variables are nonnegative.

LINGO is primarily intended for use as optimization software. The general mathematical programming or optimization model consists of the objective function, the constraints, and the decision variables. In the absence of an objective function, LINGO can also be used as an equation solver. On the other hand, multiple-objective optimization problems can only be handled by converting them into equivalent single-objective formulations.

The objective function contains the expression that should be optimized. The objective is either to maximize, as in the case of profit, yield, or efficiency, or to minimize as in the case of cost, environmental impact, or risk. Constraints are conditions that must be met by the model when finding the optimal solution. These set the limitations of the model. Such constraints include equality (e.g., conservation laws of mass and energy), inequality (e.g., lower and upper bound limits), or conditional (e.g., selection of option, if-then statements) constraints. Finally, the decision variables are those which can be controlled or modified in the system being modeled. Solving the model entails identifying the correct values of the decision variables in order to achieve the objective function within the limitations set by the constraints.

- The objective function—LINGO—is primarily an optimization software; thus, to declare the objective function, the word `max` or `min` is written followed by the equality sign and the expression that must be maximized or minimized. The end of the objective function should be signified by a semicolon ";".

- Writing a comment—To write a comment in the main display window, it must be preceded with an exclamation point "!" and it should end with a semicolon ";". Comments are written in green font when they appear in the main display window of LINGO.
- Writing constraints—Constraints and formula are written in the same way as they are written on paper. The mathematical operations carried out such as addition, subtraction, multiplication, and division make use of the symbols +, −, *, and /, respectively, while exponents are preceded with ^. Similarly, LINGO makes use of the symbols <, >, <=, >=, = to express less than, greater than, less than or equal to, greater than or equal to, and equality between expressions, respectively. However, no distinction is made between < and <=, and, > and >= when solving models in LINGO. Each constraint should end with a semicolon ";". Other functions are available and may be found in the downloadable LINGO manual.
- Declaring negative variables—Variables which can take on a negative value should be indicated by using the command @FREE(Variable_name).
- Declaring integer variables—There are two kinds of integer variables which can be declared; binary variables which can only take the value of 0 or 1 are declared using the command @BIN(Variable_name), while variables which can take any integer value are declared using the command @GIN(Variable_name).

3.4 The Optimization Model

To demonstrate how an optimization model is implemented in LINGO, two simple examples will be discussed. The first will be a simple case study which shows how to encode basic formulations, while the second case study will revisit Example 2.3 of the previous chapter.

3.4.1 Example 3.1

This first example is meant to show how simple mathematical operations are written in LINGO. A simple optimization model is given as follows:

$$\text{Objective Function: max } 10x + y + 2z \tag{3.1a}$$

Subject to:

$$10x \leq z \tag{3.1b}$$

$$x + y + z = 4 \tag{3.1c}$$

$$y \geq 0 \tag{3.1d}$$

$$z \in \{0, 1\} \tag{3.1e}$$

The above model can simply be encoded on the main display of LINGO as shown in Fig. 3.4. Note that there is no need to explicitly write the constraint corresponding to Eq. 3.1d, since by default LINGO treats all variables as being nonnegative. Figures 3.4, 3.5, 3.6, 3.7, and 3.8 illustrate some important details when coding in LINGO. These unique features deviate from the conventional way of writing optimization model components and are thus a common cause of coding errors among beginners. First, it is necessary to put an equal sign, "=", in the objective function (Fig. 3.5). The end of every line of code in LINGO must end with a semicolon ";" as shown in Fig. 3.6. In fact, two or more lines of code (e.g., constraints) can be written on the same line on screen, provided that they are separated by semicolons. The start of any remark in a model can be signified by an exclamation point, "!", while its end is still indicated by a semicolon ";" (Fig. 3.7). LINGO commands automatically appear in blue font, as shown in Fig. 3.8.

Once the model has been encoded in the main display window, it can be saved as a LINGO model file, using an appropriate filename with the filename extension ". lg4". The solution can easily be obtained by clicking on the solve button, , as shown in Fig. 3.9. Appropriate solvers are utilized depending on model type (e.g., the simplex algorithm for linear programs, or the branch-and-bound algorithm for mixed-integer linear programs).

A solution report is then generated upon pressing "Solve" as shown in Fig. 3.10 which shows the optimal values of the objective function (Fig. 3.11) and the model variables (Fig. 3.12). Additional information for sensitivity analysis is also included in the output. The report can be saved as a LINGO report, using an appropriate filename with the format indicated by the filename extension ".lgr".

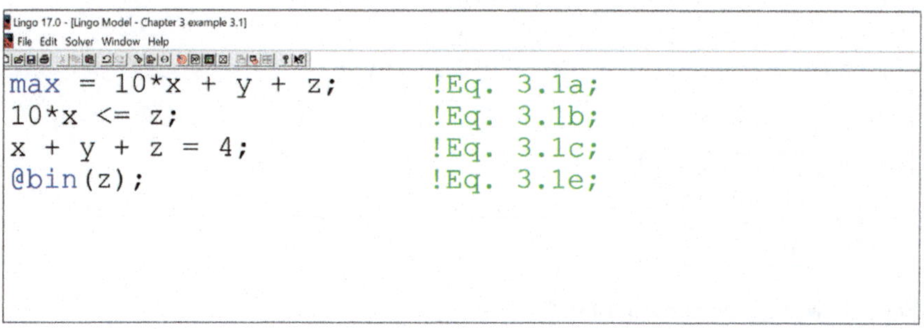

Fig. 3.4 Coding Example 3.1 in the main display window of LINGO

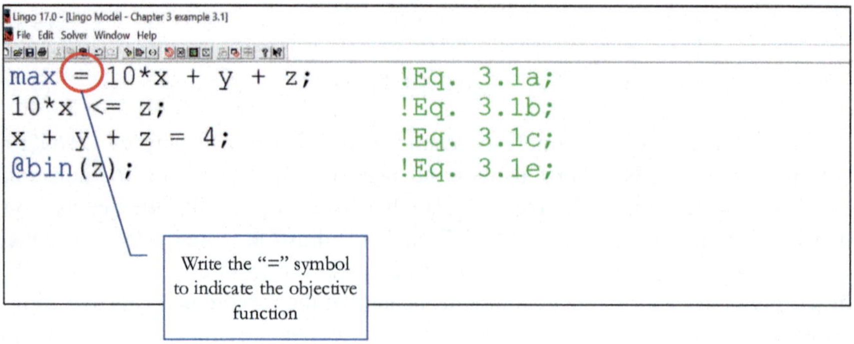

Fig. 3.5 Writing the objective function

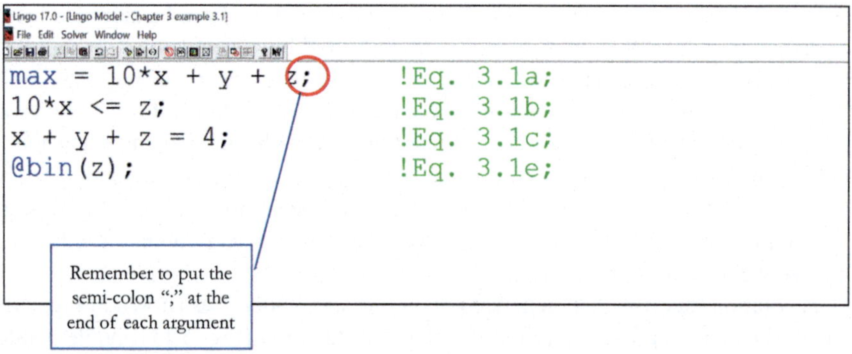

Fig. 3.6 Ending arguments in LINGO

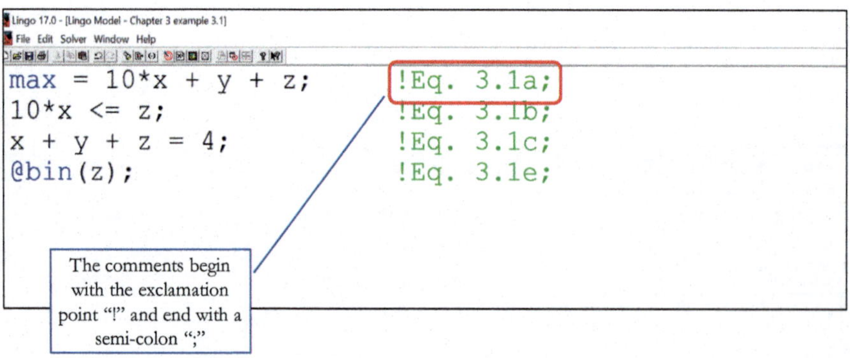

Fig. 3.7 Writing comments in LINGO

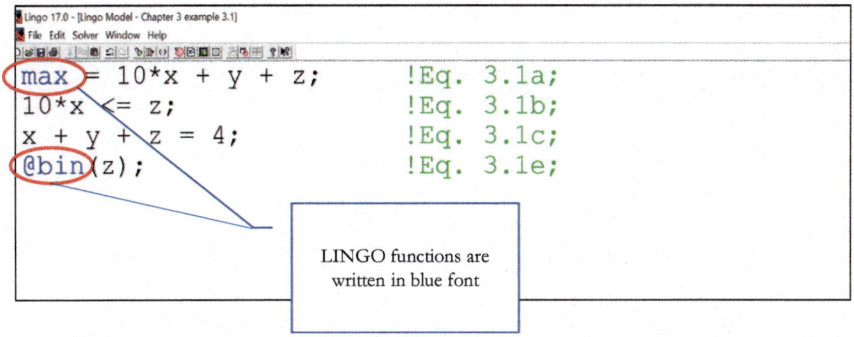

Fig. 3.8 LINGO functions

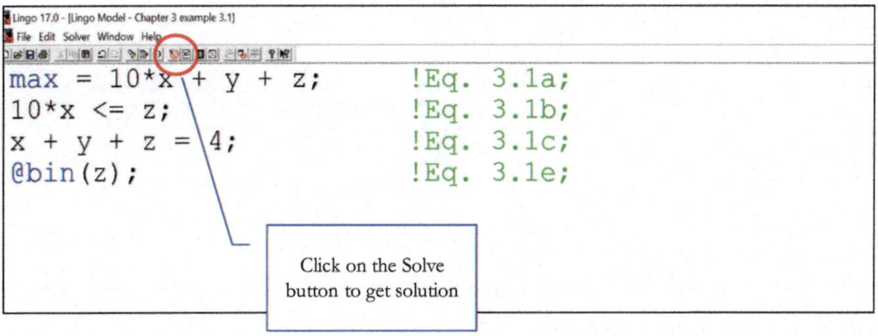

Fig. 3.9 Finding the solution of the optimization model

3.4.2 Example 3.2

This second example revisits Example 2.3 which was discussed in the previous chapter which has the objective of maximizing the gross domestic product (GDP) (Eq. 3.2a) of an economy subject to the constraints shown in Eq. 3.2b–3.2e. The hypothetical economy consists of two sectors where the decision variables f_1 and f_2 refer to the final demand of sectors 1 and 2, respectively, while decision variables x_1 and x_2, correspond to the total output or size of sectors 1 and 2. Matrix **A** is the technical coefficient matrix. Equation 3.2b–3.2e can be translated using linear algebra. The transformation is shown in Eq. 3.3a–3.3e.

$$\text{Maximize GDP} \tag{3.2a}$$

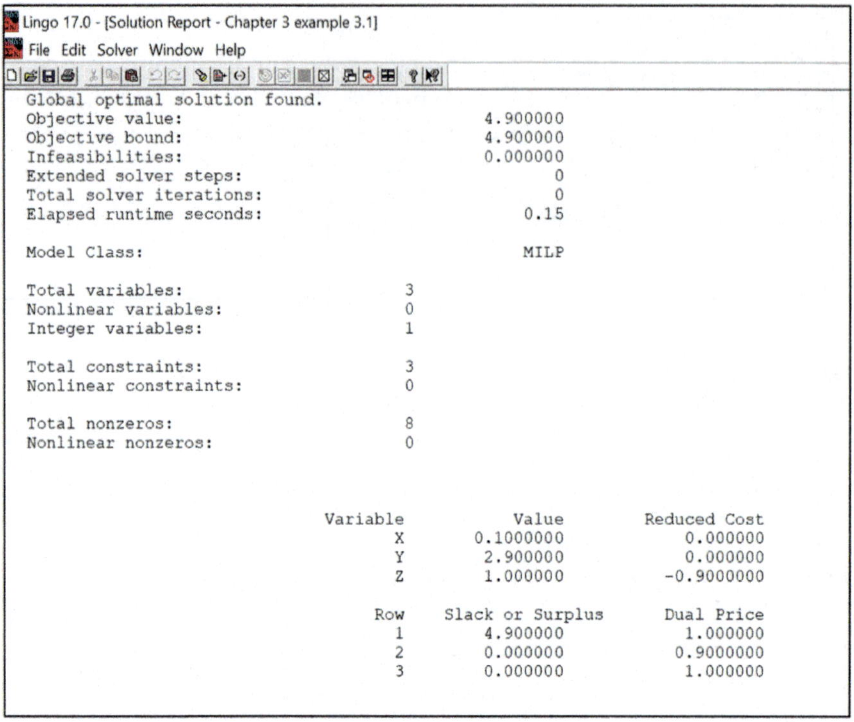
Lingo 17.0 - [Solution Report - Chapter 3 example 3.1]
File Edit Solver Window Help

```
Global optimal solution found.
Objective value:                              4.900000
Objective bound:                              4.900000
Infeasibilities:                              0.000000
Extended solver steps:                               0
Total solver iterations:                             0
Elapsed runtime seconds:                          0.15

Model Class:                                      MILP

Total variables:                    3
Nonlinear variables:                0
Integer variables:                  1

Total constraints:                  3
Nonlinear constraints:              0

Total nonzeros:                     8
Nonlinear nonzeros:                 0


                              Variable           Value        Reduced Cost
                                     X       0.1000000            0.000000
                                     Y        2.900000            0.000000
                                     Z        1.000000          -0.9000000

                                   Row    Slack or Surplus      Dual Price
                                     1        4.900000            1.000000
                                     2        0.000000           0.9000000
                                     3        0.000000            1.000000
```

Fig. 3.10 Solution report for Example 3.1

Subject to:

$$\text{GDP} = \mathbf{pf} \tag{3.2b}$$

$$\mathbf{x} = \mathbf{Ax} + \mathbf{f} \tag{3.2c}$$

$$(\mathbf{I} - \mathbf{A})\mathbf{x} = \mathbf{f} \tag{3.2d}$$

$$\mathbf{Bx} \leq \mathbf{r} \tag{3.2e}$$

$$\mathbf{A} = \begin{bmatrix} 0.15 & 0.25 \\ 0.20 & 0.05 \end{bmatrix} \tag{3.2f}$$

$$\mathbf{I} - \mathbf{A} = \begin{bmatrix} 0.85 & -0.25 \\ -0.20 & 0.95 \end{bmatrix} \tag{3.2g}$$

$$\mathbf{B} = \begin{bmatrix} 0.40 & 0.10 \\ 0.10 & 0.40 \end{bmatrix} \tag{3.2h}$$

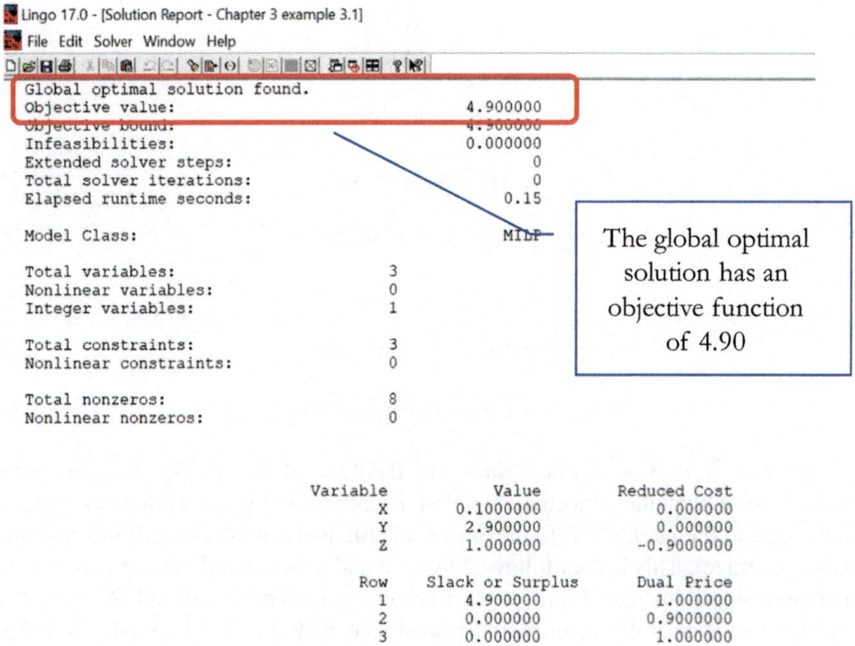

Fig. 3.11 Value of the objective function

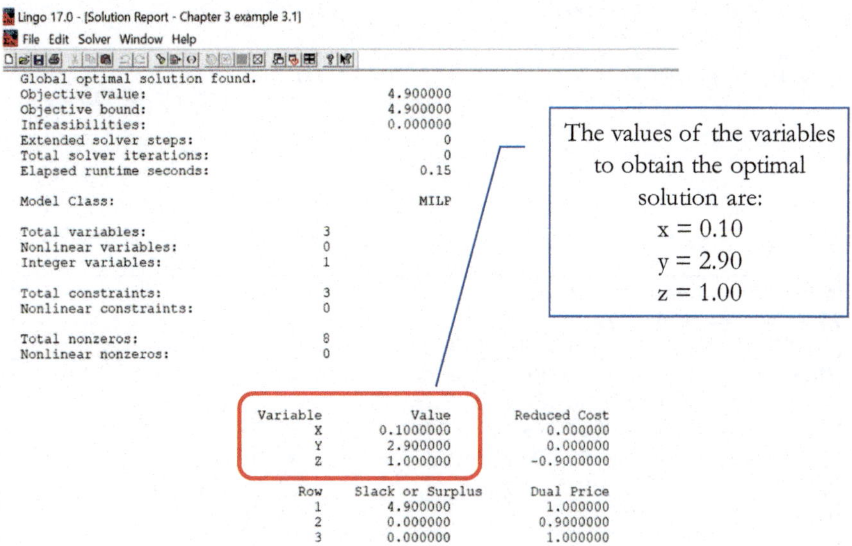

Fig. 3.12 Optimal value of model variables

$$\mathbf{r} = \begin{bmatrix} 600 \\ 900 \end{bmatrix} \tag{3.2i}$$

$$\mathbf{p} = [1 \quad 1] \tag{3.2j}$$

$$GDP = 1f_1 + 1f_2 \tag{3.3a}$$

$$f_1 = 0.85x_1 - 0.25x_2 \tag{3.3b}$$

$$f_2 = -0.20x_1 + 0.95x_2 \tag{3.3c}$$

$$0.40x_1 + 0.10x_2 \leq 600 \tag{3.3d}$$

$$0.10x_1 + 0.40x_2 \leq 900 \tag{3.3e}$$

Equation 3.3a–3.3e can be written in LINGO as shown in Fig. 3.12 and listed below. Note that the objective function is indicated by the command max, to indicate that the intention is to maximize (alternatively, min is used if the objective is to minimize). This is then followed by an equal sign (=) and the expression to be maximized (minimized). Words that are recognized as commands in LINGO appear in blue text within the mainframe as illustrated in Fig. 3.7. The model is solved simply by clicking on the solve button (Fig. 3.13).

```
max = GDP;                              !Eq. 3.2a;
GDP = f1 + f2;                          !Eq. 3.3a;
f1 = 0.85*x1 - 0.25*x2;                 !Eq. 3.3b;
f2 = -0.20*x1 + 0.95*x2;                !Eq. 3.3c;
0.40*x1 + 0.10*x2 < 600;                !Eq. 3.3d;
0.10*x1 + 0.40*x2 < 900;                !Eq. 3.3e;
```

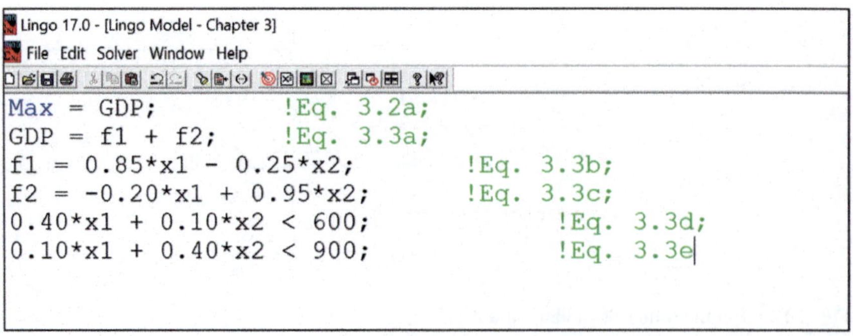

Fig. 3.13 Coding Example 3.2 in the main display window of LINGO

3.5 Set-Based Model

Direct coding of the model components is easy to learn and convenient for small problems. However, for larger models, such an approach can be tedious. Consider, for example, the prospect of manually coding an input–output system with 500 sectors; this approach will involve 500 equations, each with 500 terms on the left-hand side and another term on the right-hand side. An alternative approach is needed. Set-based models, in which model components appear symbolically instead of explicitly, are typically used when similar calculations or constraints should be implemented for objects that belong in the same group. In this way, calculations may be expressed more concisely. This form of coding is also more convenient when dealing with matrices and in developing generic optimization models, which are intended for repeated use and differ only in the number of members in defined sets. In addition, handling larger sets of data is facilitated by this approach. For example, data from spreadsheets can be cut and pasted directly into the LINGO model code to eliminate the need for manual transcription. LINGO is capable of being configured to interface directly for both data input and solution output with Excel, but the details of such capabilities are beyond the scope of this book.

For set-based coding, it is important to identify ahead of time which objects belong in the same group or set. Objects belonging in the same group are those which share the same characteristics or parameters. In Example 2.3, the two economic sectors can be grouped into one set. Each sector is characterized by the size (x_j) and the final demand (f_i). Sets may either be primitive sets or derived sets. Derived sets are sets whose objects are either a subset of a primitive set or objects derived from several primitive sets. Primitive sets which are derived from more than one set create n number of objects equivalent to all possible combinations which can be created between the objects in the primitive sets.

For set-based coding, the LINGO model may be divided into three sections—set declaration, data section, and model section.

- Set declaration—This section is meant to indicate whether certain objects should be grouped together into one set. This section begins with the command "SETS:" and ends with "ENDSETS". Set names can make use of alphanumeric symbols and the underscore "_". An example is shown below.

```
SETS:                    !beginning of Set declaration;
SETNAME1: attribute1, attribute2;       !creates the primitive
set with name "SETNAME1";
SETNAME2: attribute3;
SETNAME3  (SETNAME1,  SETNAME2):  attribute  4;  !creates  the
derived set with name "SETNAME3" with objects from "SETNAME1"
and "SETNAME2";
ENDSETS     !ends the set declaration section;
```

- Data section—This section defines the number of members in a given set and the values assigned to parameters if these are known.

```
DATA:
SETNAME1  = A1,  A2,  A3;          !the  set  SETNAME1  has  3  objects
named 1,  2 and 3;
SETNAME2 = B1, B2;         !the names of objects in a set can use
any alphanumeric symbol;
attribute1 = 10,   20,    30;  !this indicates that the value of
attribute1 is 10 for member A1, 20 for A2 and 30 for A3;
attribute2 = 5,    10,    15;
attribute3  =  100,  200;    !this  indicates  that  the  value  of
attribute3 is 100 for member B1 and 200 for B2;
attribute4 = 2,    4,     6
             3,    5,     7;
ENDDATA
```

Once the set declaration and data sections are complete, it is possible to check if the values have been assigned correctly to the model parameters by clicking on the "Solve" button. The above example generates the following information in the solution report.

```
Feasible solution found.
  Total solver iterations:                                  0
  Elapsed runtime seconds:                               0.09

  Model Class:                                          . . .

  Total variables:                  0
  Nonlinear variables:              0
  Integer variables:                0

  Total constraints:                0
  Nonlinear constraints:            0

  Total nonzeros:                   0
  Nonlinear nonzeros:               0

                                      Variable           Value
                                ATTRIBUTE1( A1)       10.00000
                                ATTRIBUTE1( A2)       20.00000
                                ATTRIBUTE1( A3)       30.00000
                                ATTRIBUTE2( A1)       5.000000
                                ATTRIBUTE2( A2)       10.00000
                                ATTRIBUTE2( A3)       15.00000
                                ATTRIBUTE3( B1)       100.0000
                                ATTRIBUTE3( B2)       200.0000
                            ATTRIBUTE4( A1, B1)       2.000000
                            ATTRIBUTE4( A1, B2)       4.000000
                            ATTRIBUTE4( A2, B1)       6.000000
                            ATTRIBUTE4( A2, B2)       3.000000
                            ATTRIBUTE4( A3, B1)       5.000000
                            ATTRIBUTE4( A3, B2)       7.000000
```

- Matrix functions—LINGO has recently included matrix functions within the platform. These include things like matrix inversion (@INVERSE(matrixA)), matrix multiplication (@MTXMUL(matrixA, matrix)) and getting the determinant of the matrix (@DETERMINANT(matrixA)) to name a few. It is important to note that the elements of the matrix have been identified early on using the set-based formulation. Matrix calculations may be performed within the CALC section of LINGO.

Take, for example, a 2 × 2 matrix (AMATRIX) whose elements, IA(*i*, *j*), consist of the interaction between "Sectors." This matrix can be declared in LINGO as given below.

```
SETS:
!This section creates the groups used in the problem;
!A single set is created, 'Sectors';
!The Amatrix set is a derived set containing elements from set
'Sectors' ;

Sectors: x;        !Sectors are defined by 3 parameters, x, f and p;
Amatrix (Sectors, Sectors): IA, IAINV; !The Amatrix is a Sector by
Sector matrix. Here it is used to house the elements of matrix (IA);
ENDSETS

DATA:
!This section contains all the known parameters of the problem
domain;
Sectors =    S1    S2;

IA =  0.85  -0.25
      -0.20 0.95;

ENDDATA
```

The inverse of matrix **A,** which now contains elements IAINV(*i*, *j*), can easily be calculated within the CALC section as follows.

```
CALC:

IAINV = @INVERSE(IA);
ENDCALC
```

Clicking the solve button for this model results in the solution report indicated below.

```
Feasible solution found.
  Infeasibilities:                               0.000000
  Total solver iterations:                              0
  Elapsed runtime seconds:                           0.10

  Model Class:                                         LP

  Total variables:                    2
  Nonlinear variables:                0
  Integer variables:                  0

  Total constraints:                  0
  Nonlinear constraints:              0

  Total nonzeros:                     0
  Nonlinear nonzeros:                 0


                               Variable           Value
                                X( S1)         0.000000
                                X( S2)         0.000000
                           IA( S1, S1)        0.8500000
                           IA( S1, S2)       -0.2500000
                           IA( S2, S1)       -0.2000000
                           IA( S2, S2)        0.9500000
                       IAINV( S1, S1)          1.254125
                        IAINV( S1, S2)        0.3300330
                        IAINV( S2, S1)        0.2640264
                       IAINV( S2, S2)          1.122112
```

- Looping functions—The two main looping functions which will be used throughout this book are @FOR and @SUM.

  @FOR is used when an equation needs to be executed repeatedly for all members of a set. The syntax is @FOR(set_name (set_index)|condition: expression). This will be discussed in more detail later.
  @SUM is used to take the sum of parameter values of objects belonging in the same set. The syntax is @SUM (set_name(set_index)|condition: expression). This will be discussed in more detail later.

Let us revisit Example 2.3. Notice that economic sectors are characterized by the total output (x) and the final demand (f). Similarly, the environmental resources are characterized by the amount of resource available (r). It is then logical to group sectors into the same set category while resources can be grouped in another.

Equation 3.2a–3.2c and 3.2e can be translated into Eqs. 3.4a–3.4d.

$$\max = \mathrm{GDP} \tag{3.4a}$$

$$\mathrm{GDP} = \sum_{i=1}^{n} p_i f_i \quad \forall i \in N \tag{3.4b}$$

$$f_i = \sum_{j=1}^{N} \mathrm{IA}_{ij} x_j \quad \forall i \in N \tag{3.4c}$$

$$\sum_{j=1}^{N} b_{kj} x_j \le r_k \quad \forall k \in Q \tag{3.4d}$$

The entire optimization for Example 2.3 is shown below.

```
SETS:
!This section creates the groups used in the problem;
!Two kinds of sets are created, the 'Sectors' and the 'Resources';
!The Amatrix and Bmatrix sets are derived sets containing elements from set
'Sectors' and 'Resources';

Sectors: x, f, P;        !Sectors are defined by 3 parameters, x, f and p;
Amatrix (Sectors, Sectors): IA; !The Amatrix is a Sector by Sector matrix.
Here it is used to house the elements of matrix (I-A);
Resources: r; !Resources are defined by parameter r which represents the
maximum resource available;
Bmatrix(Resources, Sectors): b; !The Bmatrix contains the environmental
impact of each sector on the available resources;
ENDSETS

DATA:
!This section contains all the known parameters of the problem domain;
Sectors =    S1     S2;
Resources = R1     R2;

IA =    0.85   -0.25
        -0.20  0.95;

b =     0.40   0.10
        0.10   0.40;

r =     600    900;

p = 1 1;
ENDDATA

!This section contains the optimization model of Example 2.3;

max = GDP;                                                    !Eq. 3.16;

GDP = @sum(sectors(i): p(i)*f(i));                            ! Eq. 3.17;

@for(sectors(i): f(i) = @sum(sectors(j): IA(i,j)*x(j)));      !Eq. 3.18;

@for(resources(k): @sum(sectors(j): b(k,j)*x(j)) <= r(k));    !Eq. 3.19;
```

Even if the set-based model seems longer than the original, the expression of the equations becomes convenient when one needs to deal with a system that has more than two economic sectors and two resources and when several sets of data need to be evaluated. For example, it is possible to cut and paste the necessary coefficients from spreadsheet applications directly into the data section of the code given above.

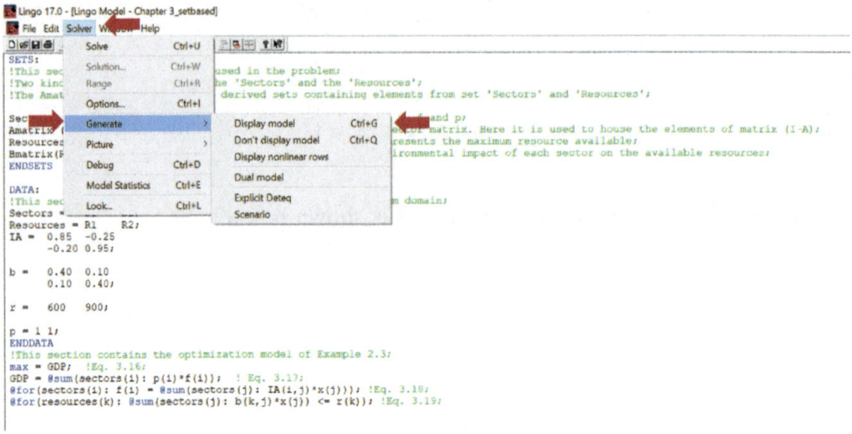

Fig. 3.14 How to display the model of set-based formulation

The entire model for the set-based formulation can easily be displayed by choosing Solver → Generate → Display Model or by simply pressing Ctrl + G as shown in Fig. 3.14.

The displayed working model is given below.

```
MODEL:
  [_1] MAX= GDP;
  [_2] GDP - F_S1 - F_S2 = 0;
  [_3] - 0.85 * X_S1 + F_S1 + 0.25 * X_S2 = 0;
  [_4] 0.2 * X_S1 - 0.95 * X_S2 + F_S2 = 0;
  [_5] 0.4 * X_S1 + 0.1 * X_S2 <= 600;
  [_6] 0.1 * X_S1 + 0.4 * X_S2 <= 900;
  END
```

Note that LINGO automatically substitutes parameters into the model.

3.6 Interpreting the Results

Once the model is completed, the solution can easily be found by clicking on the "Solve" button. The solution report should look like the one found in Fig. 3.15.

The solution report includes the optimal value of the objective function which is indicated at the top of the report as objective value. In this example, the optimal solution is 2050. It also reports model statistics which includes model class, number of variables and their categories, and number of constraints. In addition, it lists the value of all parameters and decision variables for the optimal solution. The solution report also contains the reduced cost if the dual price option is activated. The reduced cost is the number of units the coefficient of a variable should be changed in order to include it in the optimal solution. All variables which appear selected or

```
Global optimal solution found.
Objective value:                              2050.000
Infeasibilities:                              0.000000
Total solver iterations:                             4
Elapsed runtime seconds:                          0.12

Model Class:                                        LP

Total variables:                   5
Nonlinear variables:               0
Integer variables:                 0

Total constraints:                 6
Nonlinear constraints:             0

Total nonzeros:                   14
Nonlinear nonzeros:                0



                        Variable           Value        Reduced Cost
                             GDP        2050.000            0.000000
                         X( S1)         1000.000            0.000000
                         X( S2)         2000.000            0.000000
                         F( S1)         350.0000            0.000000
                         F( S2)         1700.000            0.000000
                         P( S1)         1.000000            0.000000
                         P( S2)         1.000000            0.000000
                     IA( S1, S1)       0.8500000            0.000000
                     IA( S1, S2)      -0.2500000            0.000000
                     IA( S2, S1)      -0.2000000            0.000000
                     IA( S2, S2)       0.9500000            0.000000
                         R( R1)         600.0000            0.000000
                         R( R2)         900.0000            0.000000
                      B( R1, S1)       0.4000000            0.000000
                      B( R1, S2)       0.1000000            0.000000
                      B( R2, S1)       0.1000000            0.000000
                      B( R2, S2)       0.4000000            0.000000

                             Row    Slack or Surplus      Dual Price
                               1        2050.000            1.000000
                               2        0.000000            1.000000
                               3        0.000000            1.000000
                               4        0.000000            1.000000
                               5        0.000000            1.266667
                               6        0.000000            1.433333
```

Fig. 3.15 Solution report of LINGO

activated in the optimal solution will have a reduced cost equal to zero. The solution report also contains the Slack or Surplus. The Slack or Surplus indicates how far each row is from satisfying a constraint as an equation. The dual price (also known as the shadow price) is the change in the optimal objective function value per unit of incremental change in the value of the constant of any given constraint. This feature can be used for local sensitivity analysis without having to resolve the model.

3.7 Conclusions

This chapter discusses the basics of programming in LINGO. The LINGO programming platform is relatively simple since mathematical models can easily be built using language which is similar to how formulas are written on paper. It is also possible to handle matrices and vectors using the set-based formulation in LINGO. This feature is especially important for handling large-scale models. Essential commands needed for modeling the examples outlined in this book are discussed.

References

1. Lindo Systems Inc. (2017). Available at https://lindo.com/index.php/ls-downloads?id=112:lingo-documentation&catid=82. Accessed on 26 May 2018
2. Schrage LE (2006) Optimization modeling with LINGO. Lindo System

Chapter 4
Input–Output Models of Industrial Complexes

Abstract Industrial complexes are comprised of clusters of industrial plants linked via a supply chain within close geographic proximity. Such complexes may emerge spontaneously or through deliberate planning as industrial parks. These systems when properly planned can optimize resource use and reduce waste and are known as eco-industrial parks. This chapter discusses the use of input–output models for the analysis and optimization of such industrial complexes under various conditions. An illustrative example is discussed here and is supported by LINGO model formulations.

Keywords Industrial complex · Industrial symbiosis · Industrial park

4.1 Introduction

Industrial complexes have emerged in the recent years for economic and environmental reasons. The concept of industrial ecology (IE) emerged as a framework to plan biologically inspired industrial systems that mimic the looped flows in natural ecosystems through increased recycling and reuse. One form of implementation of IE is industrial symbiosis (IS), which occurs within a special type of industrial complex known as eco-industrial parks (EIPs). The potential of input–output models to provide decision support for IE was first recognized by Duchin [5]. More recently, the related concept of the circular economy (CE) has also become popular [14]. Some governments provide incentives for firms to situate their factories in industrial parks to boost employment and, at the same time, foster collaborative practices among its occupants to maximize the benefits of industrial symbiosis. Firms co-located in an industrial complex can minimize their costs through agglomeration. Integrated waste management systems and renewable power generation are some of the services that industrial complexes offer to achieve economic and environmental benefits for the firms. Over the years, these services have expanded and industrial complexes, which focus on emission reduction, have evolved into eco-industrial parks, which geared toward waste reduction and

R. R. Tan et al., *Input–Output Models for Sustainable Industrial Systems*,
Lecture Notes in Management and Industrial Engineering,
https://doi.org/10.1007/978-981-13-1873-3_4

improving the environmental performance of firms [10]. Kalundborg is the leading example wherein firms and the local municipality exhibit symbiotic exchange of by-products for power generation and wastewater treatment, among others [9].

Growing concerns about climate change, coupled with national commitments to curb emissions (e.g., under the Paris Accord), have made it attractive for developed countries to transfer their production to developing countries, with a goal to reduce their emissions. Furthermore, developing countries have adapted the model in order to attract foreign direct investments into their economy. In addition, one of the 17 sustainable development goals highlights responsible consumption and production, which includes efficient use of resources for production [17]. International agencies such as United Nations Industrial Development Organization, World Bank Group, and Deutsche Gesellschaft für Internationale Zusammenarbeit [18] have jointly prepared an international framework for setting a standard for eco-industrial parks. This chapter discusses the use of input–output modeling in optimization for industrial complexes and a case study using LINGO.

4.2 Industrial Symbiosis for Industrial Complexes

Industrial symbiosis in industrial complexes allows firms co-located in the same industrial complex to benefit from energy, waste, and materials exchange that can reduce environmental damage and costs. Through adapting the input–output framework, the economy is translated into a physical exchange of materials and energy for production, consumption, recycling, and waste disposal. Capitalizing on the structure of the industrial system, external shocks can affect the way the system works such that changes in availability of resources can induce changes in the system's output. The role of industrial symbiosis in industrial complexes can be addressed through identifying the objective of the users, which can span from minimizing costs, maximizing gains from agglomeration activities, and minimizing environmental emissions among others.

Strategies such as strengthening initiatives in preserving natural resources, establishing an institutional system for cleaner production, and transforming industrial parks into eco-industrial parks have been identified to ensure sustainability of the expansion of industrial symbiosis in Ulsan, South Korea [13]. However, barriers to achieving industrial symbiosis within an industrial park may still exist as in the case of the Philippines, wherein the lack of awareness of industrial symbiosis is the main reason why firms co-located within an industrial park do not maximize the opportunity to collaborate among each other [3].

Optimization models play a key role in planning out exchanges within industrial complexes to achieve optimal network design [2]. Furthermore, these models can be designed to account for bi-level decision-making wherein the regulating body is treated as the leader and firms are treated as followers whose decisions are influenced by the decision of the leader [1].

The interdependent nature between firms within an industrial complex leads to vulnerability to disruptions and their ripple effects. This vulnerability becomes more pronounced in tightly integrated systems that are optimized from the standpoint of resource use and pollution generation. Using network analysis, Chopra and Khanna [4] were able to identify vulnerable points in industrial complexes that can critically affect the system. In addition, quantifying resilience through index metrics can produce insights on each component in the system [6]. With the looming effects of climate change, industrial complexes should move toward climate-proofing industrial networks and become imperative through decision-making support tools that will factor in climate change impacts for developing effective policies [15]. This is vital in assuring the sustainability of the industrial complex. While numerous models have been developed to consider the case of industrial complexes, this chapter will consider an optimization model as discussed in Tan et al. (2016) that maximizes the overall level of satisfaction of fuzzy goals during crisis conditions, wherein the normal condition is presented using a standard enterprise input–output model defined as follows:

$$\mathbf{Ax} = \mathbf{y} \tag{4.1}$$

$$\mathbf{Bx} = \mathbf{z} \tag{4.2}$$

where $\mathbf{A}$ is the process matrix, $\mathbf{x}$ is the normal state production level vector, $\mathbf{y}$ is the vector of net outflows from the industrial complex, $\mathbf{B}$ is the resource matrix, and $\mathbf{z}$ is the vector of resource inputs in the industrial complex. Each column of the process matrix, $\mathbf{A}$, denotes the proportion of industrial inputs per unit of product output of a given plant. Positive values denote outputs, while negative values denote inputs. Similarly, each column of the resource matrix, $\mathbf{B}$, denotes the proportion of resource input per unit of product output of a plant. Resource inputs are denoted using absolute values. The production vector, $\mathbf{x}$, and the resource inputs vector, $\mathbf{z}$, have nonnegative elements, while the net outflows vector, $\mathbf{y}$, exhibits nonnegative elements unless the material is sourced externally, in which case the industrial complex imports the material, hence, a negative element in $\mathbf{y}$.

Under crisis conditions, the optimization model is specified as:

$$\max \lambda \tag{4.3}$$

subject to:

$$\mathbf{Ax}^* = y^* \tag{4.4}$$

$$\mathbf{Bx}^* = z^* \tag{4.5}$$

$$\mathbf{x}^* \geq \lambda(\mathbf{x} - \mathbf{x_L}) + \mathbf{x_L} \tag{4.6}$$

$$\mathbf{y}^* \geq \lambda(\mathbf{y} - \mathbf{y_L}) + \mathbf{y_L} \quad (4.7)$$

$$\mathbf{z}^* \geq \lambda(\mathbf{z_L} - \mathbf{z_U}) + \mathbf{z_U} \quad (4.8)$$

where λ is the overall level of satisfaction fuzzy constraints, **x*** is the production level vector under crisis condition, **y*** is the net outflows vector under crisis condition, **z*** is the resource inputs vector under crisis condition, $\mathbf{x_L}$ is the exogenously defined lower limit of production level that is acceptable to the firms, $\mathbf{y_L}$ is the exogenously defined lower limit of acceptable net outflows, $\mathbf{z_L}$ is the lower limit of the estimated available resource input resulting from the crisis, and $\mathbf{z_U}$ is the upper limit of the estimated available resource input resulting from the crisis. It should be noted that λ will assume values from 0 to 1, wherein a value of 1 indicates complete satisfaction of the fuzzy constraints, 0 indicates nonsatisfaction, and values in between indicate partial satisfaction.

This model provides an optimal solution for scenarios wherein an introduction of a perturbation in the level of production, net outflows, and resource inputs available to the industrial complex resulting from an external shock such as drought, power shortage, and flooding. Other disruptions to the system may be brought about by changes in regulation, failure of interconnections among individual firms, and disruptions to individual firm operations [7].

4.3 Case Study 4.1

This example is based on the example discussed in Tan et al. (2016) wherein the case of an aluminum industrial complex with an objective of maximizing an overall level of satisfaction of fuzzy goals such as resource consumption, production level, and net final product outflow under crisis conditions. This system considers eight process units to ultimately produce three products, namely shape cast aluminum, extruded aluminum, and cold-rolled aluminum. Table 4.1 presents the process matrix (**A**) in the first eight rows and the resource matrix (**B**) in the last row.

Under normal conditions, Table 4.2 presents the baseline flow of materials and energy for the aluminum complex for processing 1 t/h of aluminum ingot. In particular, it shows that to produce 0.738 t/h of shape cast aluminum, 0.228 t/h of extruded aluminum, and 0.034 t/h of cold-rolled aluminum, the system needs a total of 970.74 t/h of water for all the processes, with a large proportion amounting to 958.20 t/h being attributable to power generation.

Figure 4.1 shows the visual representation of the flows within the aluminum industrial complex with the shape cast aluminum, extruded aluminum, and cold-rolled aluminum shaded.

A hypothetical drought scenario is considered such that an estimate of 80–85% of water is available for the aluminum complex. This will yield potential changes in the level of production of shape cast aluminum, extruded aluminum, and cold-rolled aluminum. Based on such information, the management determines an acceptable level of reduction in production for the three different products. More specifically, it

Table 4.1 Process and resource matrix of aluminum complex

	Power generation	Mining	Primary smelting	Recycling	Shape casting	Extrusion	Hot rolling	Cold rolling
Electricity (kWh)	1.00	−0.40	−15,785	−115	−4.00	−590	−265	−350
Bauxite (t)	0	1	−5.044	0	0	0	0	0
Aluminum ingot (t)	0	0	1	1	−1.1	−1.5	−1.2	0
Shape cast aluminum (t)	0	0	0	0	1	0	0	0
Extruded aluminum (t)	0	0	0	0	0	1	0	0
Hot-rolled aluminum (t)	0	0	0	0	0	0	1	−1.2
Cold-rolled aluminum (t)	0	0	0	−1	0.1	0.5	0.2	0.2
Scrap aluminum (t)	0	0	0	−1	0.1	0.5	0.2	0.2
Water (t)	0.06	0.205	6.922	0.96	5.45	1.18	1.352	0.13

Table 4.2 Baseline flows for aluminum complex

	Power generation	Mining	Primary smelting	Recycling	Shape casting	Extrusion	Hot rolling	Cold rolling	Net output
Electricity (kW)	15,971	−2.018	−15,785	−23.32	−2.952	−134.50	−10.81	−11.9	0.00
Bauxite (t/h)	0	5.044	−5.044	0	0	0	0	0	0.00
Aluminum ingot (t/h)	0	0	1	0.203	−0.812	−0.342	−0.049	0	0.00
Shape cast aluminum (t/h)	0	0	0	0	0.738	0	0	0	0.738
Extruded aluminum (t/h)	0	0	0	0	0	0.228	0	0	0.228
Hot-rolled aluminum (t/h)	0	0	0	0	0	0	0.041	−0.041	0.00
Cold-rolled aluminum (t/h)	0	0	0	0	0	0	0	0.034	0.034
Scrap aluminum (t/h)	0	0	0	−0.203	0.074	0.114	0.008	0.007	0.00
Water (t/h)	958.20	1.034	6.922	0.195	4.022	0.269	0.055	0.005	970.74

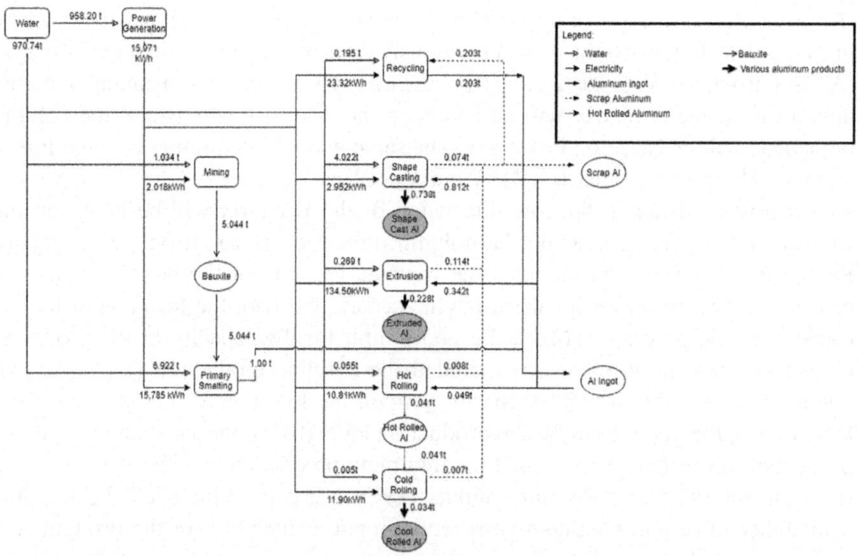

Fig. 4.1 Material and energy flows in the aluminum complex for the baseline scenario (flowrate per hour)

finds that a minimum of 70% from the baseline production is acceptable for shape cast aluminum, 80% for extruded aluminum, and 90% for cold-rolled aluminum. Table 4.3 presents a summary of the fuzzy limits of estimated water resource availability and acceptable levels of production considering the drought.

To setup the model in LINGO using set-based model, the sets should be declared initially. For this case study, the sets are declared as follows:

```
Sets:

Processes: x, xnew, LLx, ULx;
Products: y, ynew, LLy, ULy, lambday, alpha;
Envtlimpact: g, gnew, LLg, ULg, lambdag;
Amatrix (Products, Processes): a;
Bmatrix(Envtlimpact, Processes): b;

Endsets
```

This paragraph of code is explained as follows. There are variables that are defined as to have the dimension of process units (`Processes`), production

Table 4.3 Fuzzy limits of estimated water resource availability and acceptable levels of production for drought scenario

	Percentage of baseline (%)	Lower limit	Upper limit
Water resource availability (t/h)	80–85	776.60	825.13
Shape cast aluminum (t/h)	70–100	0.517	0.738
Extruded aluminum (t/h)	80–100	0.182	0.228
Cold-rolled aluminum (t/h)	90–100	0.031	0.034

(`Products`), and resource inputs (`Envtlimpact`). In addition, matrices may be introduced with dimensions of sets previously declared such that the process matrix (**A**) is introduced as `Amatrix (Products, Processes)` meaning the row dimension of the **A** matrix will be based on the data in Products and the column dimension will be based on Processes. The same way of interpreting the next line is applied. The resource matrix (**B**) is introduced as `Bmatrix (Envtlimpact, Processes)` meaning the row dimension of the **B** matrix will be based on the data in `Envtlimpact`, while the column dimension is based on `Processes`. Processes (`Processes`) included the baseline capacity of the processes (`x`), the capacity of the processes for the drought scenario (`xnew`), the lower limit for the capacity of the processes (`LLx`), the upper limit for the capacity of the processes (`ULx`). Production (`Products`) included the baseline production level (`y`), production level for the drought scenario (`ynew`), the lower limit for the production level (`LLy`), the upper limit for the production level (`ULy`), the satisfaction level for production levels (`lambday`), and the minimum production level acceptable to the management (`alpha`). Resource inputs (`Envtlimpact`) included the baseline availability of resource inputs (`g`), resource input availability for the drought scenario (`gnew`), the lower limit for resource input availability under drought scenario (`LLg`), the upper limit for resource input availability under drought scenario (`ULg`), and the satisfaction level for resource input availability (`lambdag`).

After declaring the sets to be used for the model, the data wherein the sets will be based on should be encoded as well. The next block of code accomplishes this.

```
Data:
Processes = Powergen    Mining       Psmelting   Recycling   Scasting
      Extrusion   Hotrolling  Coldrolling;
Products =  electricity bauxite      alumingot   scastalum   exalum
      hotalum     coldalum     scrapalum;
Envtlimpact = water;



a =

1      -0.4   -15785  -115   -4     -590   -265   -350
0      1      -5.044  0      0      0      0      0
0      0      1       1      -1.1   -1.5   -1.2   0
0      0      0       0      1      0      0      0
0      0      0       0      0      1      0      0
0      0      0       0      0      0      1      -1.2
0      0      0       0      0      0      0      1
0      0      0       -1     0.1    0.5    0.2    0.2
;

b =
0.06   0.205 6.922 0.96   5.45   1.18   1.352 0.135
;

y = 0 0 0 0.738 0.228 0 0.034 0;

alpha = 0 0 0 .7 .8 0 .9 0;

Enddata
```

The following data are introduced into the LINGO: the process units considered (`Processes`), the products (`Products`), the resource inputs (`Envtlimpact`), the contents of the **A** matrix (`a`), the contents of the **B** matrix (`b`), the baseline production level (`y`), and the minimum production level acceptable to the management (`alpha`).

The eight process units in Table 4.1 are declared under `Processes` as power generation (`Powergen`), mining (`Mining`), primary smelting (`Psmelting`), recycling (`Recycling`), shapecasting (`Scasting`), extrusion (`Extrusion`), hot rolling (`Hotrolling`), and cold rolling (`Coldrolling`).

The eight products in Table 4.1 are declared under `Products` as electricity (`electricity`), bauxite (`bauxite`), aluminum ingots (`alumingot`), shape cast aluminum (`scastalum`), extruded aluminum (`exalum`), hot-rolled aluminum (`hotalum`), cold-rolled aluminum (`coldalum`), and scrap aluminum (`scrapalum`).

Resource input considered in Table 4.1 is declared under `Envtlimpact` as water (`water`).

Given that the data declared under `Processes`, `Products`, and `Envtlimpact`, `a` will serve as the entry for `Amatrix` with a dimension of 8×8, `b` will serve as the entry for `Bmatrix` with a dimension of 1×8, `y` will serve as the entry for `y` which is the baseline level of production with a dimension of 1×8, and `alpha` will serve as the entry for `alpha` which is the minimum production level acceptable to the management with a dimension of 1×8. These serve as the exogenous variables in the system.

The next blocks of code will specify the optimizationnmodel presented in Eqs. 4.3–4.8.

First, declare that the objective is to maximize the variable `lambda`.

```
max = lambda;
```

The next block declares the assumptions that have been made. Recall that the process matrix may contain negative values to denote that the entries are inputs and the production vector may contain negative values if the industrial complex imports products externally. These are addressed through affixing an `@free` before the matrix data.

Table 4.4 Optimal level of production and resource input under drought scenario

	Optimal level	Percentage of baseline (%)	Lower limit	Upper limit
Water resource availability (t/h)	806.96	83.13	776.60	825.13
Shape cast aluminum (t/h)	0.599	81.17	0.517	0.738
Extruded aluminum (t/h)	0.199	87.28	0.182	0.228
Cold-rolled aluminum (t/h)	0.032	94.12	0.031	0.034

```
!matrix A and vector y can contain negative numbers;
@for(amatrix(i,j): @free(a(i,j)));
@for(products(i): @free(y(i)));
```

The next block calculates the solution for the baseline scenario as declared in Eqs. 4.1 and 4.2.

```
!Calculation of base case ;
@for(products(i): y(i) = @sum(processes(j): a(i,j)*x(j)));
@for(envtlimpact(k): g(k) = @sum(processes(j): b(k,j)*x(j)));
```

The next block declares the limits under crisis conditions as given in Table 4.4. It should be noted that for the production level, we have declared `alpha` as the minimum production level acceptable to the management, which determines the lower limit of the production vector. Since it has different values for different products, we multiply the values for `alpha(i)` and `y(i)`. Furthermore, the upper limits for `products` and `processes` are equal to the baseline. In addition, given that there is only one entry for `envtlimpact`, which is water, we can do away with simply multiplying the entries in Table 4.3 pertaining to water resource availability to the baseline value.

```
!Limits for crisis scenario;
@for(products(i): LLy(i) = alpha(i)*y(i));          !limit for y;
@for(products(i): ULy(i) = y(i));
@for(processes(j): LLx(j) = 0.80*x(j));             !limit for process
capacity;
@for(processes(j): ULx(j) = x(j));
@for(envtlimpact(k): ULg(k) = 0.85*970.73);         !limit for water;
@for(envtlimpact(k): LLg(k) = 0.80*970.73);
```

Given the limits defined in the previous block, the new production levels and resource inputs can now be computed using the block of code below.

```
!Crisis scenario;
@for(products(i): ynew(i) = @sum(processes(j): a(i,j)*xnew(j)));
@for(envtlimpact(k): gnew(k) = @sum(processes(j): b(k,j)*xnew(j)));
```

The satisfaction levels for production level and resource inputs can also be computed as a transformation of Eqs. 4.7 and 4.8.

```
!satisfaction levels;
@for(products(i): lambday(i)*(ULy(i) - LLy(i)) = (ynew(i) - LLy(i)));
@for(envtlimpact(k): lambdag(k) = (ULg(k) - gnew(k))/ (ULg(k) - LLg(k)));
```

The final block of constraints requires that `lambda` is less than or equal to the individual levels of satisfaction of the product goals, as expressed in the first three lines. Note that `lambday(4)` represents the satisfaction level for `scastalum`, which is the fourth element of the `lambday` vector, `lambday(5)` represents the satisfaction level for `exalum`, and `lambday(7)` represents the satisfaction level for `coldroll`. Similarly, lambda should also be less than the value of `lambdag (k)`, which pertains to the satisfaction level of resource inputs.

```
lambda <= lambday(4);
lambda <= lambday(5);
lambda <= lambday(7);
@for(envtlimpact(k): lambdag(k) >= lambda);
```

Once the model has been set up completely, click the Solve button located at the toolbar to find the optimal solution.

The solution report will yield a global optimal solution with $\lambda = 0.3742214$. This means that there is a partial satisfaction of the fuzzy goals set for the drought scenario considered.

```
Global optimal solution found.
  Objective value:                              0.3742214
  Infeasibilities:                               0.000000
  Total solver iterations:                              0
  Elapsed runtime seconds:                           0.12

  Model Class:                                         LP

  Total variables:                   60
  Nonlinear variables:                0
  Integer variables:                  0

  Total constraints:                 48
  Nonlinear constraints:              0

  Total nonzeros:                   130
  Nonlinear nonzeros:                 0
```

In addition, the new production levels and water resource requirement are also solved as shown in Table 4.4 and highlighted in the solution report below.

Variable	Value	Reduced Cost
LAMBDA	0.3742214	0.000000
X(POWERGEN)	15970.52	0.000000
X(MINING)	5.044000	0.000000
X(PSMELTING)	1.000000	0.000000
X(RECYCLING)	0.2027600	0.000000
X(SCASTING)	0.7380000	0.000000
X(EXTRUSION)	0.2280000	0.000000
X(HOTROLLING)	0.4080000E-01	0.000000
X(COLDROLLING)	0.3400000E-01	0.000000
XNEW(POWERGEN)	13277.04	0.000000
XNEW(MINING)	4.190502	0.000000
XNEW(PSMELTING)	0.8307895	0.000000
XNEW(RECYCLING)	0.1737013	0.000000
XNEW(SCASTING)	0.5994526	0.000000
XNEW(EXTRUSION)	0.1994645	0.000000
XNEW(HOTROLLING)	0.3824682E-01	0.000000
XNEW(COLDROLLING)	0.3187235E-01	0.000000
LLX(POWERGEN)	12776.42	0.000000
LLX(MINING)	4.035200	0.000000
LLX(PSMELTING)	0.8000000	0.000000
LLX(RECYCLING)	0.1622080	0.000000
LLX(SCASTING)	0.5904000	0.000000
LLX(EXTRUSION)	0.1824000	0.000000
LLX(HOTROLLING)	0.3264000E-01	0.000000
LLX(COLDROLLING)	0.2720000E-01	0.000000
ULX(POWERGEN)	15970.52	0.000000
ULX(MINING)	5.044000	0.000000
ULX(PSMELTING)	1.000000	0.000000
ULX(RECYCLING)	0.2027600	0.000000
ULX(SCASTING)	0.7380000	0.000000
ULX(EXTRUSION)	0.2280000	0.000000
ULX(HOTROLLING)	0.4080000E-01	0.000000
ULX(COLDROLLING)	0.3400000E-01	0.000000
Y(ELECTRICITY)	0.000000	0.000000
Y(BAUXITE)	0.000000	0.000000
Y(ALUMINGOT)	0.000000	0.000000
Y(SCASTALUM)	0.7380000	0.000000
Y(EXALUM)	0.2280000	0.000000
Y(HOTALUM)	0.000000	0.000000
Y(COLDALUM)	0.3400000E-01	0.000000
Y(SCRAPALUM)	0.000000	0.000000
LAMBDAY(ELECTRICITY)	0.000000	0.000000
LAMBDAY(BAUXITE)	0.000000	0.000000
LAMBDAY(ALUMINGOT)	0.000000	0.000000
LAMBDAY(SCASTALUM)	0.3742214	0.000000
LAMBDAY(EXALUM)	0.3742214	0.000000

```
          LAMBDAY( HOTALUM)        0.000000            0.000000
         LAMBDAY( COLDALUM)        0.3742214           0.000000
        LAMBDAY( SCRAPALUM)        0.000000            0.000000
          LLY( ELECTRICITY)        0.000000            0.000000
              LLY( BAUXITE)        0.000000            0.000000
            LLY( ALUMINGOT)        0.000000            0.000000
            LLY( SCASTALUM)        0.5166000           0.000000
               LLY( EXALUM)        0.1824000           0.000000
              LLY( HOTALUM)        0.000000            0.000000
             LLY( COLDALUM)        0.3060000E-01       0.000000
            LLY( SCRAPALUM)        0.000000            0.000000
          ULY( ELECTRICITY)        0.000000            0.000000
              ULY( BAUXITE)        0.000000            0.000000
            ULY( ALUMINGOT)        0.000000            0.000000
            ULY( SCASTALUM)        0.7380000           0.000000
               ULY( EXALUM)        0.2280000           0.000000
              ULY( HOTALUM)        0.000000            0.000000
             ULY( COLDALUM)        0.3400000E-01       0.000000
            ULY( SCRAPALUM)        0.000000            0.000000
         YNEW( ELECTRICITY)        0.000000            0.000000
             YNEW( BAUXITE)        0.000000            0.000000
           YNEW( ALUMINGOT)        0.000000            0.000000
           YNEW( SCASTALUM)        0.5994526           0.000000
              YNEW( EXALUM)        0.1994645           0.000000
             YNEW( HOTALUM)        0.000000            0.000000
            YNEW( COLDALUM)        0.3187235E-01       0.000000
           YNEW( SCRAPALUM)        0.000000            0.000000
        ALPHA( ELECTRICITY)        0.000000            0.000000
            ALPHA( BAUXITE)        0.000000            0.000000
          ALPHA( ALUMINGOT)        0.000000            0.000000
          ALPHA( SCASTALUM)        0.7000000           0.000000
             ALPHA( EXALUM)        0.8000000           0.000000
            ALPHA( HOTALUM)        0.000000            0.000000
           ALPHA( COLDALUM)        0.9000000           0.000000
          ALPHA( SCRAPALUM)        0.000000            0.000000
                  G( WATER)        970.7327            0.000000
               GNEW( WATER)        806.9571            0.000000
                LLG( WATER)        776.5840            0.000000
                ULG( WATER)        825.1205            0.000000
            LAMBDAG( WATER)        0.3742214           0.000000
     A( ELECTRICITY, POWERGEN)     1.000000            0.000000
       A( ELECTRICITY, MINING)    -0.4000000           0.000000
    A( ELECTRICITY, PSMELTING)    -15785.00            0.000000
    A( ELECTRICITY, RECYCLING)    -115.0000            0.000000
     A( ELECTRICITY, SCASTING)    -4.000000            0.000000
    A( ELECTRICITY, EXTRUSION)    -590.0000            0.000000
   A( ELECTRICITY, HOTROLLING)    -265.0000            0.000000
  A( ELECTRICITY, COLDROLLING)    -350.0000            0.000000
         A( BAUXITE, POWERGEN)     0.000000            0.000000
           A( BAUXITE, MINING)     1.000000            0.000000
        A( BAUXITE, PSMELTING)    -5.044000            0.000000
        A( BAUXITE, RECYCLING)     0.000000            0.000000
         A( BAUXITE, SCASTING)     0.000000            0.000000
        A( BAUXITE, EXTRUSION)     0.000000            0.000000
       A( BAUXITE, HOTROLLING)     0.000000            0.000000
      A( BAUXITE, COLDROLLING)     0.000000            0.000000
       A( ALUMINGOT, POWERGEN)     0.000000            0.000000
         A( ALUMINGOT, MINING)     0.000000            0.000000
      A( ALUMINGOT, PSMELTING)     1.000000            0.000000
      A( ALUMINGOT, RECYCLING)     1.000000            0.000000
       A( ALUMINGOT, SCASTING)    -1.100000            0.000000
      A( ALUMINGOT, EXTRUSION)    -1.500000            0.000000
     A( ALUMINGOT, HOTROLLING)    -1.200000            0.000000
    A( ALUMINGOT, COLDROLLING)     0.000000            0.000000
       A( SCASTALUM, POWERGEN)     0.000000            0.000000
         A( SCASTALUM, MINING)     0.000000            0.000000
      A( SCASTALUM, PSMELTING)     0.000000            0.000000
      A( SCASTALUM, RECYCLING)     0.000000            0.000000
       A( SCASTALUM, SCASTING)     1.000000            0.000000
      A( SCASTALUM, EXTRUSION)     0.000000            0.000000
     A( SCASTALUM, HOTROLLING)     0.000000            0.000000
    A( SCASTALUM, COLDROLLING)     0.000000            0.000000
          A( EXALUM, POWERGEN)     0.000000            0.000000
```

```
      A( EXALUM, MINING)        0.000000        0.000000
    A( EXALUM, PSMELTING)       0.000000        0.000000
    A( EXALUM, RECYCLING)       0.000000        0.000000
     A( EXALUM, SCASTING)       0.000000        0.000000
    A( EXALUM, EXTRUSION)       1.000000        0.000000
   A( EXALUM, HOTROLLING)       0.000000        0.000000
  A( EXALUM, COLDROLLING)       0.000000        0.000000
    A( HOTALUM, POWERGEN)       0.000000        0.000000
      A( HOTALUM, MINING)       0.000000        0.000000
   A( HOTALUM, PSMELTING)       0.000000        0.000000
   A( HOTALUM, RECYCLING)       0.000000        0.000000
    A( HOTALUM, SCASTING)       0.000000        0.000000
   A( HOTALUM, EXTRUSION)       0.000000        0.000000
  A( HOTALUM, HOTROLLING)       1.000000        0.000000
 A( HOTALUM, COLDROLLING)      -1.200000        0.000000
   A( COLDALUM, POWERGEN)       0.000000        0.000000
     A( COLDALUM, MINING)       0.000000        0.000000
  A( COLDALUM, PSMELTING)       0.000000        0.000000
  A( COLDALUM, RECYCLING)       0.000000        0.000000
   A( COLDALUM, SCASTING)       0.000000        0.000000
  A( COLDALUM, EXTRUSION)       0.000000        0.000000
 A( COLDALUM, HOTROLLING)       0.000000        0.000000
A( COLDALUM, COLDROLLING)       1.000000        0.000000
  A( SCRAPALUM, POWERGEN)       0.000000        0.000000
    A( SCRAPALUM, MINING)       0.000000        0.000000
 A( SCRAPALUM, PSMELTING)       0.000000        0.000000
 A( SCRAPALUM, RECYCLING)      -1.000000        0.000000
  A( SCRAPALUM, SCASTING)      0.1000000        0.000000
 A( SCRAPALUM, EXTRUSION)      0.5000000        0.000000
A( SCRAPALUM, HOTROLLING)      0.2000000        0.000000
A( SCRAPALUM, COLDROLLING)     0.2000000        0.000000
      B( WATER, POWERGEN)      0.6000000E-01    0.000000
        B( WATER, MINING)      0.2050000        0.000000
     B( WATER, PSMELTING)       6.922000        0.000000
     B( WATER, RECYCLING)      0.9600000        0.000000
      B( WATER, SCASTING)       5.450000        0.000000
     B( WATER, EXTRUSION)       1.180000        0.000000
    B( WATER, HOTROLLING)       1.352000        0.000000
   B( WATER, COLDROLLING)      0.1350000        0.000000
```

The resulting network for the disrupted system is as illustrated in Fig. 4.2.

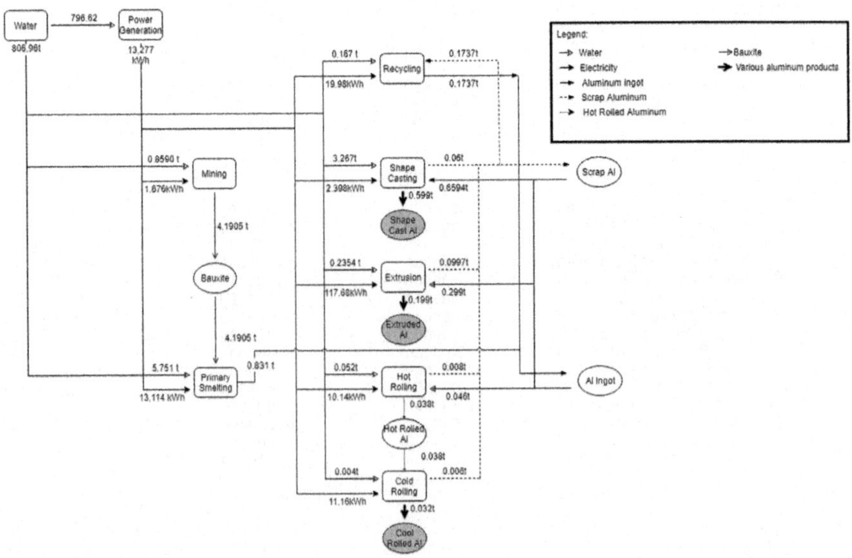

Fig. 4.2 Process flows in disrupted aluminum complex

4.4 Conclusions

There is a potential for applying mathematical optimization models such as the example discussed to include the emerging issues of using Industry 4.0 tools to encourage decision makers in adapting industrial symbiosis [16]. Initial work on complexity modeling of vulnerability within industrial complexes has been done to account for topological structure [12] and failures within the network [11]. In addition to promoting industrial symbiosis among firms in industrial complexes, firms can also harness sustainable practices through their supply chain [8]. This topic will be discussed further in Chap. 9.

References

1. Aviso KB, Tan RR, Culaba AB, Cruz JB (2010) Bi-level fuzzy optimization approach for water exchange in eco-industrial parks. Process Saf Environ Prot 88:31–40
2. Aviso KB (2014) Design of robust water exchange networks for eco-industrial symbiosis. Process Saf Environ Prot 92:160–170
3. Bacudio LR, Benjamin MFD, Eusebio RCP, Holaysan SAK, Promentilla MAB, Yu KDS, Aviso KB (2016) Analyzing barriers to implementing industrial symbiosis networks using DEMATEL. Sustain Prod Consumption 7:57–65
4. Chopra SS, Khanna V (2014) Understanding resilience in industrial symbiosis networks: insights from network analysis. J Environ Manage 141:86–94
5. Duchin F (1992) Industrial input-output analysis: implications for industrial ecology. Proc Nat Acad Sci U S A 89:851–855
6. Fraccascia L, Giannoccaro I, Albino V (2017) Rethinking resilience in industrial symbiosis: conceptualization and measurements. Ecol Econ 137:148–162
7. Kuznetsova E, Zio E, Farel R (2016) A methodological framework for eco-industrial park design and optimization. J Clean Prod 126:308–324
8. Leigh M, Li X (2015) Industrial ecology, industrial symbiosis and supply chain environmental sustainability: a case study of a large UK distributor. J Clean Prod 106:632–643
9. Jacobsen NB (2006) Industrial symbiosis in Kalundborg, Denmark: a quantitative assessment of economic and environmental aspects. J Ind Ecol 10(1–2):239–255
10. Kechichian E, Jeong MH (2016) Mainstreaming eco-industrial parks: conclusions from the eco-industrial park 2015 event in Seoul (English). World Bank Group, Washington, DC
11. Li B, Xiang P, Hu M, Zhang C, Dong L (2017) The vulnerability of industrial symbiosis: a case study of Qijiang Industrial Park, China. J Clean Prod 157:267–277
12. Li X, Xiao R (2017) Analyzing network topological characteristics of eco-industrial parks from the perspective of resilience: a case study. Ecol Ind 74:403–413
13. Park H-S, Rene ER, Choi S-M, Chiu ASF (2008) Strategies for sustainable development of industrial park in Ulsan, South Korea—from spontaneous evolution to systematic expansion of industrial symbiosis. J Environ Manage 87:1–13
14. Prieto-Sandoval V, Jaca C, Ormazabal M (2018) Towards a consensus on the circular economy. J Clean Prod 179:605–615
15. Tan RR, Aviso KB, Chiu ASF, Promentilla MAB, Razon LF, Tseng ML, Yu KDS (2017) Towards "climate-proof" industrial networks. Resour Conserv Recycl 127:244–245
16. Tseng ML, Tan RR, Chiu ASF, Chien CF, Kuo TC (2018) Circular economy meets industry 4.0: can big data drive industrial symbiosis? Resour Conserv Recycl 131:146–147

17. United Nations (n.d.). Goal 12: ensure sustainable consumption and production pattern. Accessed from http://www.un.org/sustainabledevelopment/sustainable-consumption-production/ on 15 Mar 2018
18. United Nations Industrial Development Organization, World Bank Group, Deutsche Gesellschaft für Internationale Zusammenarbeit (2017) An international framework for eco-industrial parks. World Bank, Washington, DC

Chapter 5
Input–Output Models of Infrastructure Systems

Abstract Infrastructure systems consisting of mutually interdependent components are common in modern urban settings. Such systems are constantly at risk of cascading failure from various triggering events or perturbations. Consequently, such perturbations cause indirect ripple effects to other interdependent systems, which can propagate through the network as inoperability. Inoperability input–output models are extensions of input–output models that can represent the flow of inoperability through such critical infrastructure systems. Such models are discussed in this chapter, supported by illustrative examples and corresponding LINGO code.

Keywords Critical infrastructure systems · Perturbation · Ripple effects
Inoperability · Inoperability input–output models

5.1 Introduction

An infrastructure system is a network of systems that provides the essential services or goods to support a nation's economy and quality of life. For example, the following systems such as telecommunications, electric power systems, natural gas and oil, banking and finance, transportation, water supply systems, government services, and emergency services are typically considered as critical infrastructure systems (CISs) . Such systems are considered as critical since the destruction of the infrastructure or just a disruption in their operation would have a debilitating impact. In addition, CIS is inherently complex in its structure, characterized by interdependencies among its components which may include physical flows of goods, flows of information. This issue of dependency becomes more important when a failure in one system spreads beyond itself to cause appreciable impact on the other infrastructures and cascades further to other systems to aggravate the situation. One classic example of such cascading failure is the 1998 Galaxy IV satellite failure that caused disruption not only on the telecommunication services but also to transportation and emergency services. Such "inoperability" or the

R. R. Tan et al., *Input–Output Models for Sustainable Industrial Systems*,
Lecture Notes in Management and Industrial Engineering,
https://doi.org/10.1007/978-981-13-1873-3_5

failure of a system to deliver its intended function and how the inoperability in one system cascades through other infrastructure systems must be understood to manage the risk in CIS and to ensure the continuity of the goods and services provided by these systems.

This chapter shows how the input–output framework, using the concept of inoperability, introduced in Haimes and Jiang [1] can elucidate the "ripple effects" attributed to cascading failure through interdependent infrastructure systems. This inoperability input–output modeling (IIM) framework is also used as the tool for optimal allocation of operability to each component in order to minimize damage caused by disruptions. This model can be implemented in the optimization software LINGO.

5.2 Inoperability Input–Output Model (IIM)

The IIM framework is based on economic input–output (I–O) analysis which was based on a well-known theory of market equilibrium (Leontief [4]. This framework describes linkages among economic sectors via linear equations and is a sufficiently well-established tool for such purposes as economic analysis. Fundamental principles and many of the potential applications can be found in modern textbooks [5]. While conventional I–O models measure physical or monetary flows of goods, the IIM framework focuses on the spread of operability degradation in the infrastructure system. Haimes and Jiang [1] propose inoperability as a dimensionless quantity that can be used as a normalized index of risk in infrastructure systems. Inoperability can also be described as the degree of failure of a system to deliver its intended output due to internal failures or external perturbations. Inoperability assumes a value of 0 for systems in normal state, fractional values for partially dysfunctional systems, and 1 for systems in a state of complete failure. Subsequent work refined the definition of inoperability as a fractional loss relative to a nominal desired state, which has enabled the calibration of IIM models from government economic statistics [6]. Alternative approaches for the calibration of IIM models have also been proposed, for example, via fuzzy set theory and expert judgment [7]. The methodology has been applied to the analysis of various disruptive events, such as intentional attacks [6] and loss of key economic or natural resources (Khanna and Bakshi [3]. A brief description of an optimization-based extension of IIM was also given by Haimes and Jiang [1] and developed further in a subsequent paper [2]. Recent studies also proposed a linear programming model to evaluate the capacity of systems to absorb exogenous shocks [9] and optimally allocate scarce resources [10].

The basic formulation of the IIM [1] is:

$$(\mathbf{I}-\mathbf{A}^{*})\mathbf{q} = \mathbf{c}^{*} \tag{5.1}$$

where **I** is the identity matrix, $\mathbf{A}^*$ is the square interdependency matrix, **q** is the inoperability vector, and $\mathbf{c}^*$ is the perturbation vector. Each element in $\mathbf{A}^*$ gives the degree of influence or interdependency of one component on another. For feasible systems, this equation may be rearranged to give:

$$\mathbf{q} = (\mathbf{I} - \mathbf{A}^*)^{-1}\mathbf{c}^* \tag{5.2}$$

Equation (5.2) gives the total system inoperability, **q**, as a function of the perturbation $\mathbf{c}^*$. The inverse $(\mathbf{I} - \mathbf{A}^*)^{-1}$ has amplification properties, such that each element of **q** is larger than the corresponding element in $\mathbf{c}^*$. The amplification simply means that total losses (**q**) are greater than the direct impacts ($\mathbf{c}^*$) due to additional indirect effects resulting from system topology as described by $\mathbf{A}^*$.

Both Eqs. (5.1) and (5.2) have zero degrees of freedom, thus allowing a unique solution to be identified for a given system. By introducing new assumptions, additional degrees of freedom may be obtained to allow for system optimization, resulting in the linear programming (LP) model:

$$\text{minimize } \mathbf{w}^{\mathrm{T}}\mathbf{q} \tag{5.3a}$$

subject to:

$$\mathbf{q} = \min\left\{(\mathbf{I} - \mathbf{A}^*)^{-1}\mathbf{c}^*, 1\right\} \tag{5.3b}$$

$$\mathbf{c}^*_{\mathrm{L}} \leq \mathbf{c}^* \leq \mathbf{c}^*_{\mathrm{U}} \tag{5.3c}$$

$$0 \leq q_j \leq 1 \forall j \tag{5.3d}$$

where **w** is the weight vector that quantifies the importance of each infrastructure system component, and $\mathbf{c}^*_{\mathrm{L}}$ and $\mathbf{c}^*_{\mathrm{U}}$ are the lower and upper limits of perturbation $\mathbf{c}^*$. For the perturbed infrastructure $i = k$, we have $c^*_{k,\mathrm{L}} = c^*_{k,\mathrm{U}} = c^*_k$. For the infrastructure system components not subject to direct disruption ($i \neq k$), it is assumed that some $c^*_{i,\mathrm{L}}$ can assume small negative values due to the presence of spare capacity. Thus, the spare capacity may be used to introduce negative perturbations in the system so as to offset anticipated losses from ripple effects. It should be noted that such negative perturbation levels have not been discussed in IIM literature, although in principle there can be physical basis for their existence. Hence, this optimization model minimizes the total weighted inoperability of the system for a given perturbation from an adverse event using such interventions. Additional system constraints can be added to extend this into a mixed-integer linear programming (MILP) formulation, as described by Tan [8].

5.3 Case Study 5.1

This case study is based on the example presented in Haimes and Jiang [1]. The system is comprised of:

- Component 1—Power infrastructure (e.g., power plant and transmission grid)
- Component 2—Transportation infrastructure (e.g., road and rail networks)
- Component 3—Health infrastructure (e.g., hospitals and clinics)
- Component 4—Commercial infrastructure (e.g., grocery stores and retail outlets).

The coefficients of the interdependency matrix $\mathbf{A}^*$ are given in Table 5.1, while the system is illustrated in Fig. 5.1. Note that the potential flow of inoperability within the system is in reverse; it emanates from an external triggering perturbation and cascades through the interconnected network components. The coefficients may be interpreted as in the following example. Complete failure of the power infrastructure ($j = 1$), i.e., 100% inoperability, will result in 40% inoperability in the transportation infrastructure ($i = 2$) and 100% inoperability of the health ($i = 3$) and commercial infrastructure ($i = 4$). Likewise, a complete failure of the transportation ($j = 2$) will result in 90% inoperability in the power infrastructure ($i = 1$); i.e., the system can only provide 10% of its intended capacity. On the other hand, any disruption on operation in the health infrastructure ($j = 3$) and commercial infrastructure ($j = 4$) will not cause any significant impact on each other and to the other infrastructures (e.g., power and transportation).

Table 5.2 describes overall inoperability transmission matrix analogous to the Leontief inverse matrix. The coefficients in the diagonal matrix ($i = j$) can be interpreted as a multiplier to determine how much an internal failure in the ith infrastructure is being amplified due to the presence of interdependencies. Likewise, the other coefficients ($i \neq j$) describe how much of the inoperability that cascades through the system from an external failure or perturbation c^* in the jth infrastructure is transmitted to the ith infrastructure, taking into account both the direct and ripple effects.

Table 5.1 Inoperability interdependency matrix ($\mathbf{A}^*$) for the infrastructure system [1]

	Power infrastructure $j = 1$	Transport infrastructure $j = 2$	Health infrastructure $j = 3$	Commercial infrastructure $j = 4$
Power ($i = 1$)	0	0.90	0	0
Transport ($i = 2$)	0.40	0	0	0
Health ($i = 3$)	1	0.80	0	0
Commercial ($i = 4$)	1	0.90	0	0

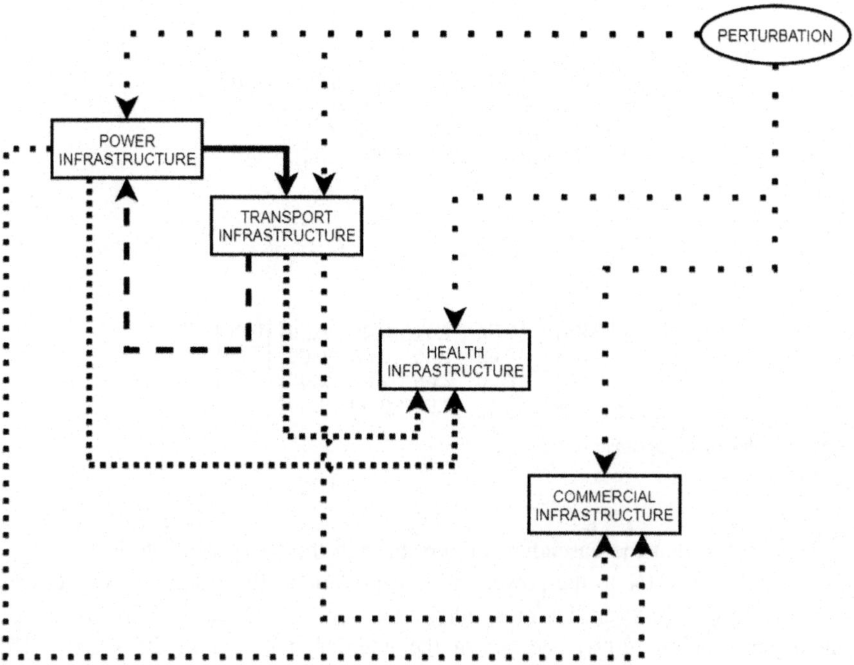

Fig. 5.1 Schematic diagram of interdependency between critical infrastructure systems

Table 5.2 Overall inoperability transmission matrix ($(\mathbf{I}-\mathbf{A}^{*})^{-1}$) for the infrastructure system

	$j = 1$	$j = 2$	$j = 3$	$j = 4$
$i = 1$	1.56	1.41	0	0
$i = 2$	0.63	1.56	0	0
$i = 3$	2.06	2.66	1.0	0
$i = 4$	2.13	2.81	0	1.0

Suppose a natural disaster hits the area and renders the transportation system partially inoperable by 20% of its intended function. Due to this disruption, many workers such as those in health infrastructures are not able to get to work, while delivery trucks are not able to arrive on time to provide service to the commercial infrastructure. Using Eq (5.2), we can quantify the risk of such failure in transportation infrastructure due to external perturbations as follows:

Inoperability of power infrastructure

Perturbation in transport infrastructure

$$q = \begin{bmatrix} 0.28 \\ 0.31 \\ 0.53 \\ 0.56 \end{bmatrix} = \begin{bmatrix} 1.56 & 1.41 & 0.00 & 0.00 \\ 0.63 & 1.56 & 0.00 & 0.00 \\ 2.06 & 2.66 & 1.00 & 0.00 \\ 2.13 & 2.81 & 0.00 & 1.00 \end{bmatrix} \begin{bmatrix} 0.00 \\ 0.20 \\ 0.00 \\ 0.00 \end{bmatrix}$$

Inoperability of transportation infrastructure

Inoperability of health infrastructure

$$q = \begin{bmatrix} 0.28 \\ 0.31 \\ 0.53 \\ 0.56 \end{bmatrix} = \begin{bmatrix} 1.56 & 1.41 & 0.00 & 0.00 \\ 0.63 & 1.56 & 0.00 & 0.00 \\ 2.06 & 2.66 & 1.00 & 0.00 \\ 2.13 & 2.81 & 0.00 & 1.00 \end{bmatrix} \begin{bmatrix} 0.00 \\ 0.20 \\ 0.00 \\ 0.00 \end{bmatrix}$$

Inoperability of commercial infrastructure

This means that the cascading failure through the system results in an inoperability of 28 and 31% to the power and transportation infrastructures, respectively. Note that inoperability in the transportation sector is amplified by 1.56 times of the initial perturbation (20%) because of the interdependencies in the system. This illustrates that the inoperability propagated by the transportation sector to other transportation-dependent infrastructure leads to higher level of inoperability and causes additional inoperability in the transportation sector itself. Likewise, this causes systems failure that leads to 53 and 56% inoperability for the health and commercial infrastructure, respectively. In practice, this amplification effect is a general feature of real infrastructure networks.

Using this model, it is also possible to evaluate the impact of the disaster of varying intensity on the operability of the system [1]. For example, consider $h \times 100\%$ is the intensity of the disaster that causes the initial perturbation to the transportation sector; i.e., $h \times 100\%$ of the operability of the transportation system is degraded due to this disaster. Using the given Leontief inverse matrix in Table 5.2, we have the following:

$$q = \begin{bmatrix} q_1 \\ q_2 \\ q_3 \\ q_4 \end{bmatrix} = \begin{bmatrix} 1.56 & 1.41 & 0.00 & 0.00 \\ 0.63 & 1.56 & 0.00 & 0.00 \\ 2.06 & 2.66 & 1.00 & 0.00 \\ 2.13 & 2.81 & 0.00 & 1.00 \end{bmatrix} \begin{bmatrix} 0 \\ h \\ 0 \\ 0 \end{bmatrix}$$

Note that this system of linear equations has one degree of freedom with additional inequality constraints such that the final inoperability and h assume to have values between 0 and 1 inclusive (i.e., $0 \leq q_j \leq 1; 0 \leq h \leq 1$). Thus, the upper limit of the initial perturbation ($\mathbf{c}_U^*$) for the transportation system wherein the solution exists is 0.356 satisfying these constraints. At this initial perturbation ($h = 0.356$), the final inoperability of the commercial infrastructure (q_4) becomes 1.

This means that the commercial infrastructure is the first to be out of commission at this intensity ($h = 0.356$), whereas the final inoperability of the power, transportation, and health infrastructures are 0.50, 0.55, and 0.94, respectively.

This problem can also be coded and solved in LINGO as follows (Fig. 5.2).

```
MAX = h;

q1 = 1.41*h; !Eq. 5.1;
q2 = 1.56*h; !Eq. 5.1;
q3 = 2.66*h; !Eq. 5.1;
q4 = 2.81*h; !Eq. 5.1;

h <= 1;
q1 <=1;
q2 <=1;
q3 <=1;
q4 <=1;
```

The objective function is to maximize the allowable perturbation in the transportation infrastructure which can be handled by the entire system before at least one sector becomes fully inoperable (i.e., inoperability cannot exceed 1.0). Note that there is no need to include nonnegativity constraints in the model since LINGO assumes that all model variables are nonnegative.

A similar strategy can be used to determine when the next failure will occur in the system. By deactivating the constraint for the inoperable sector as shown in Fig. 5.3 and further maximizing h. The optimal solution results in an increased initial perturbation to the transportation system equivalent to 0.376. When this occurs, both the health (q_3) and commercial infrastructure (q_4) become out of commission; i.e., the final inoperability of these two subsystems is 1.0. Such cascading failure also results to a final inoperability of the power infrastructure of $q_1 = 0.53$, and the final inoperability of transportation infrastructure of $q_2 = 0.59$ from the initial inoperability of the transportation infrastructure of $h = 0.376$. This

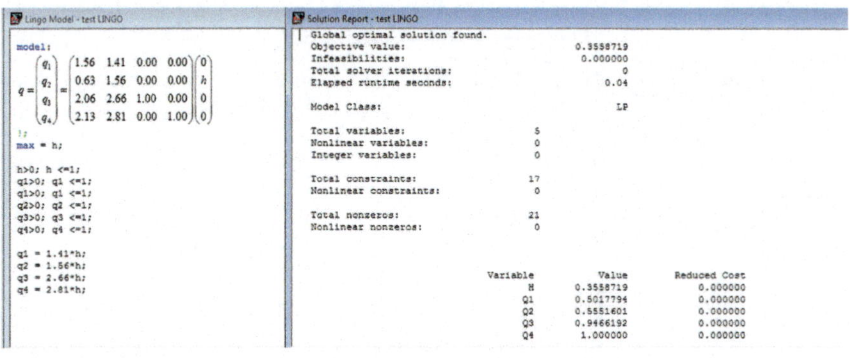

Fig. 5.2 Screenshot of LINGO model and solution report

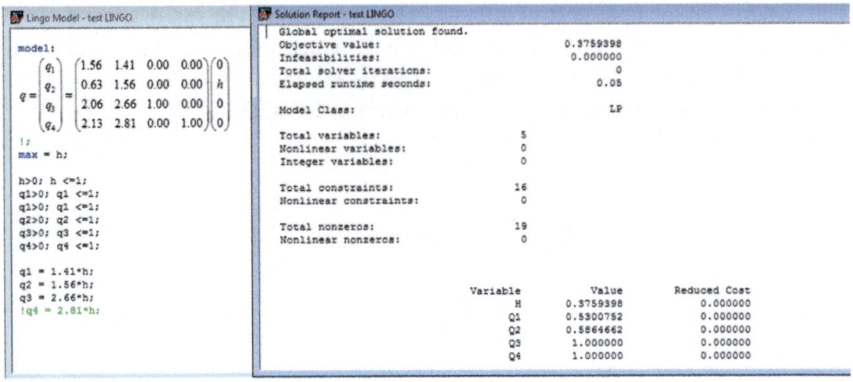

Fig. 5.3 Screenshot of LINGO model when commercial infrastructure is fully inoperable

problem can be coded similarly in LINGO wherein the equation to compute the final inoperability of the commercial infrastructure (q_4) is no longer included from the set of constraints since it is already out of commission at $h = 0.356$. Notice that we can also use Eq. (5.3b) as one of the constraints to ensure this optimization problem has a solution.

Likewise, succeeding complete failures can be determined sequentially by deactivating the constraints of the already inoperable sectors (health and commercial). The next failure is observed to occur in the transportation infrastructure ($q_2 = 1.0$) with an h perturbation of 0.641. This corresponds in the final inoperability of the power sector of $q_1 = 0.90$ as shown in Fig. 5.4.

Consider now a scenario wherein the transportation infrastructure is partially inoperable by 20% but it is assumed that there is excess capacity in the power infrastructure which is normally left unutilized. Such spare capacity may be activated to offset the system disruptions. The objective is to allocate reduced infrastructure capacity to give the minimum level of total system-level dysfunction. Note that it is assumed that the total system-level dysfunction (i.e., aggregated

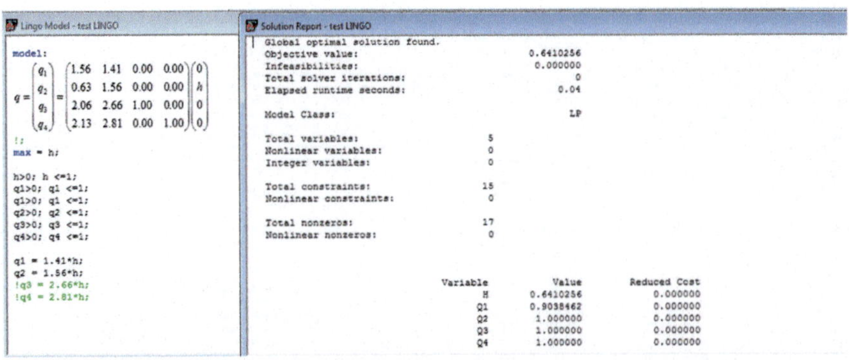

Fig. 5.4 Screenshot of LINGO model when transport infrastructure is fully inoperable

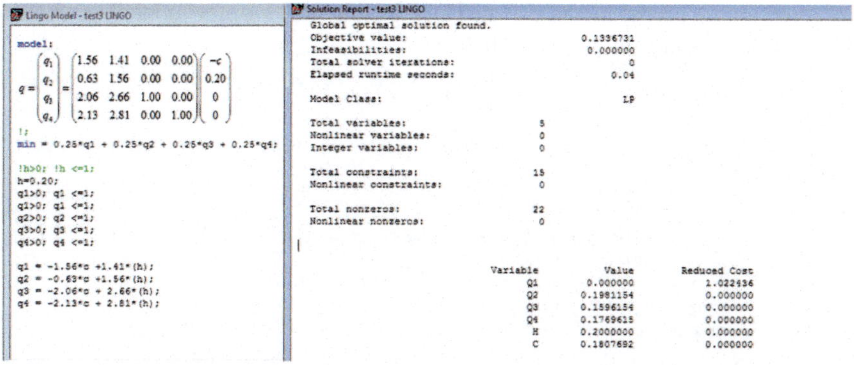

Fig. 5.5 Screenshot of LINGO model when power infrastructure has excess capacity

inoperability) takes into account both direct and indirect impacts of the failure due to the interdependencies among the components. This optimization problem as described in Eq. 5.3a can be coded and solved in LINGO as follows (Fig. 5.5):

Similarly, if the model was translated using set-based coding, the result is as shown below. Parameter IA (i, j) corresponds to the result of the operation $(\mathbf{I} - \mathbf{A})$ where $\mathbf{I}$ is an identity matrix and $\mathbf{A}$ is the technical coefficient matrix given in Table 5.1. For the case study given, the weight (wt) associated with the different sectors is all equal to 0.25.

```
SETS:
SECTORS:c, q, wt ;
AMATRIX (SECTORS, SECTORS): IA;

ENDSETS

DATA:

SECTORS = POWER   TRANSPO     HEALTH      COMMERCIAL;

IA =        1     -0.90       0           0
            -0.40 1           0           0
            -1    -0.80       1           0
            -1    -0.90       0           1;

C =         ,     0.20,       0,          0;
wt =        0.25  0.25        0.25        0.25;

ENDDATA
MIN = @SUM(SECTORS(J): WT(J)*Q(J));

@FREE(C(1));
C(1) > -1; C(1) < 1;
@FOR(SECTORS(I): C(I)  = @SUM(SECTORS(J): IA(I,J)*Q(J)));
```

Alternatively, if the model was displayed using the "Display model" function of LINGO, the results are as shown below.

```
MODEL:
[_1] MIN= 0.25 * Q_POWER + 0.25 * Q_TRANSPO + 0.25 * Q_HEALTH + 0.25 *
Q_COMMERCIAL ;
[_2] C_POWER >= - 1;
[_3] C_POWER <= 1;
[_4] C_POWER - Q_POWER + 0.9 * Q_TRANSPO = 0;
[_5] 0.4 * Q_POWER - Q_TRANSPO = - 0.2;
[_6] Q_POWER + 0.8 * Q_TRANSPO - Q_HEALTH = 0;
[_7] Q_POWER + 0.9 * Q_TRANSPO - Q_COMMERCIAL = 0;
@FREE( C_POWER);
END
```

The following solution report is also generated from LINGO:

```
Global optimal solution found.
Objective value:                                  0.5400000
Infeasibilities:                                   0.000000
Total solver iterations:                                  0
Elapsed runtime seconds:                               0.11

Model Class:                                             LP

Total variables:                          5
Nonlinear variables:                      0
Integer variables:                        0

Total constraints:                        7
Nonlinear constraints:                    0

Total nonzeros:                          17
Nonlinear nonzeros:                       0
```

Variable	Value	Reduced Cost
C(POWER)	-0.1800000	0.000000
C(TRANSPO)	0.2000000	0.000000
C(HEALTH)	0.000000	0.000000
C(COMMERCIAL)	0.000000	0.000000
Q(POWER)	0.000000	1.020000
Q(TRANSPO)	0.2000000	0.000000
Q(HEALTH)	0.1600000	0.000000
Q(COMMERCIAL)	0.1800000	0.000000
WT(POWER)	0.250000	0.000000
WT(TRANSPO)	0.250000	0.000000
WT(HEALTH)	0.250000	0.000000
WT(COMMERCIAL)	0.250000	0.000000
IA(POWER, POWER)	1.000000	0.000000
IA(POWER, TRANSPO)	-0.9000000	0.000000
IA(POWER, HEALTH)	0.000000	0.000000
IA(POWER, COMMERCIAL)	0.000000	0.000000
IA(TRANSPO, POWER)	-0.4000000	0.000000
IA(TRANSPO, TRANSPO)	1.000000	0.000000
IA(TRANSPO, HEALTH)	0.000000	0.000000
IA(TRANSPO, COMMERCIAL)	0.000000	0.000000
IA(HEALTH, POWER)	-1.000000	0.000000
IA(HEALTH, TRANSPO)	-0.8000000	0.000000
IA(HEALTH, HEALTH)	1.000000	0.000000
IA(HEALTH, COMMERCIAL)	0.000000	0.000000
IA(COMMERCIAL, POWER)	-1.000000	0.000000
IA(COMMERCIAL, TRANSPO)	-0.9000000	0.000000
IA(COMMERCIAL, HEALTH)	0.000000	0.000000
IA(COMMERCIAL, COMMERCIAL)	1.000000	0.000000

Row	Slack or Surplus	Dual Price
1	0.1350000	-1.000000
2	0.8200000	0.000000
3	1.180000	0.000000
4	0.000000	0.000000
5	0.000000	0.675000
6	0.000000	0.250000
7	0.000000	0.250000

From this solution report, the values of perturbations and inoperability for the four infrastructure system components in the optimal solution were obtained as summarized in Table 5.3. This illustrates how excess capacity in electricity generation can be activated during the crisis to mitigate inoperability in other key system components. The modeling framework thus allows the limited system capacity to be allocated in order to minimize system-wide losses incurred as ripple effects cascade through interdependent system components.

Table 5.3 Perturbation and inoperability of infrastructure system

	Perturbation	Inoperability
Power infrastructure	−0.18	0.00
Transportation infrastructure	+0.20	0.20
Health infrastructure	0	0.16
Commercial infrastructure	0	0.18

5.3 Conclusions

The Leontief input–output framework has been extended in the form of IIM to model the flow of inoperability in infrastructure systems with interdependent components. This methodology allows ripple effects triggered by disruptive events, such as natural disasters or malicious attacks, to be predicted so as to give insights on proper preventive or protection measures. By comparison, conventional risk management approaches may apply to much focus on direct, immediately observable impacts. IIM is an excellent example of how the input–output framework extends readily to physical networked systems.

References

1. Haimes YY, Jiang P (2001) Leontief-based model of risk in complex interconnected infrastructures. J Infrastruct Syst 7:1–12
2. Jiang P, Haimes YY (2004) Risk management for Leontief-based interdependent systems. Risk Anal 24:1215–1229
3. Khanna V, Bakshi BR (2009) Modeling the risks to complex industrial networks due to loss of natural capital. In: IEEE international symposium on sustainable systems and technology, ISSST'09 in Cooperation with 2009 IEEE international symposium on technology and society, 18–20 May 2009, Tempe, Arizona, USA. Art no 515677
4. Leontief W (1936) Quantitative input and output relations in the economic system of the United States. Rev Econ Stat 18:105–125
5. Miller RE, Blair PD (2009) Input-output analysis. Foundations and extensions, 2nd edn. Cambridge University Press, Cambridge
6. Santos JR, Haimes YY (2004) Modeling the demand reduction input-output (I-O) inoperability due to terrorism of interconnected infrastructures. Risk Anal 24:1437–1451
7. Setola R, De Porcellinis S, Sforna S (2009) Critical infrastructure dependency assessment using the input–output inoperability model. Int J Crit Infrastruct Prot 2:170–178
8. Tan RR (2011) A general source-sink model with inoperability constraints for robust energy sector planning. Appl Energy 88:3759–3764
9. Tan RR, Aviso KB, Promentilla MAB, Solis FD, Yu KDS, Santos JR (2015) Shock absorption index for inoperability input-output models. Econ Syst Res 27:43–59
10. Tan RR, Aviso KB, Promentilla MAB, Yu KDS, Santos JR (2015) A fuzzy linear programming model for allocation of inoperability in economic sectors due to loss of natural resource inputs. DLSU Bus Econ Rev 24:1–12

Chapter 6
Input–Output Optimization Models for Supply Chains

Abstract Input–output models have the ability to reflect supply chain linkages in industrial networks. The vulnerability of such networks to external perturbations can also be shown within input–output framework. In this chapter, an extension of the input–output model for determining how to optimally allocate levels of production during a transient crisis is discussed. A simple didactic example is solved first, followed by a more complex case study involving climate-induced disruption. Both examples are accompanied by LINGO codes.

Keywords Supply chain · Fuzzy linear programming · Simple additive weighting

6.1 Introduction

Input–output models can be used to represent supply chain networks in entire economic systems. The transaction entries within input–output tables are direct reflections of supply chain linkages. Thus, input–output models can be used to capture interdependencies among component units (i.e., economic sectors) and thus assess ripple effects of disruptions in supply chains. This feature is essential to enable the systematic planning of sustainable and resilient systems.

Examples of some applications in the literature include the assessment of the effect of pollination losses and energy shortage [5], as well as the effect of earthquakes and tsunamis [6]. In addition to the analysis of the effects of both natural and man-made disruptions, input–output models have also been extended via mathematical programming to allow optimal responses to abnormal conditions to be determined directly from model output. A notable early example is a multiple-objective linear programming (MOLP) model developed by Kananen et al. [4], which was used to prescribe optimal reallocation of outputs in Finland in response to labor and energy shortage scenarios. Jiang and Haimes [3] also proposed a similar modeling approach for risk management in input–output systems. Such capabilities are essential for ensuring that indirect effects are properly taken into account.

R. R. Tan et al., *Input–Output Models for Sustainable Industrial Systems*,
Lecture Notes in Management and Industrial Engineering,
https://doi.org/10.1007/978-981-13-1873-3_6

In this chapter, a fuzzy linear programming (FLP) model for allocating production capacity during a crisis is described; the model uses the input–output framework to account for interdependencies and supply chain linkages [9]. The model is also linked to the environment via additional linear equations to account for the use of natural resources [1]. The chapter is organized as follows. The derivation of the model is first discussed in the next section. Then, a brief didactic case study is used to illustrate the use of the model. A second case study is also solved to show the model's effectiveness for larger, more realistic problems. Both case studies are also accompanied by the actual program code.

6.2 Model Development

The development of the model in this chapter begins with the basic input–output model, augmented with additional equations to account for the flow of natural resources:

$$(\mathbf{I}-\mathbf{A})\mathbf{x} = \mathbf{y} \tag{6.1}$$

$$\mathbf{Bx} = \mathbf{z} \tag{6.2}$$

where **I** is the identity matrix, **A** is the technical coefficient matrix, **x** is the total output vector, **y** is the final demand vector, **B** is the natural resource coefficient matrix, and **z** is the natural resource consumption matrix. As usual, Eq. 6.1 may be inverted as follows:

$$\mathbf{x} = (\mathbf{I}-\mathbf{A})^{-1}\mathbf{y} \tag{6.3}$$

Combining Eqs. 6.2 and 6.3 gives:

$$\mathbf{B}(\mathbf{I}-\mathbf{A})^{-1}\mathbf{y} = \mathbf{z} \tag{6.4}$$

The amount of natural resources consumed by the economic system, **z**, in order to produce final output, **y**, is given in compact form by Eq. 6.4. Changes in **y** and **z** are hence related as follows:

$$\mathbf{B}(\mathbf{I}-\mathbf{A})^{-1}\delta\mathbf{y} = \delta\mathbf{z} \tag{6.5}$$

Equation 6.5 can be used to determine how the loss of natural resource inputs, δ**z**, affects the economic output, δ**y**. In a typical realistic system, **y** has more rows than **z**, so that Eq. 6.5 has positive degrees of freedom. It is then possible to implement optimization. Next, Eq. 6.5 may be modified so that disruptions are normalized into dimensionless form. We define the fractional loss of final demand as:

$$\mathbf{u} = \mathbf{D}\delta\mathbf{y} \tag{6.6}$$

$$\mathbf{D} = (\text{diag}(\mathbf{y}))^{-1} \tag{6.7}$$

Likewise, we define fractional loss of natural resource as:

$$\mathbf{v} = \mathbf{E}\delta\mathbf{z} \tag{6.8}$$

$$\mathbf{E} = (\text{diag}(\mathbf{z}))^{-1} \tag{6.9}$$

Equation 6.5 is then manipulated to give:

$$\mathbf{EB}(\mathbf{I}-\mathbf{A})^{-1}\left(\mathbf{D}^{-1}\mathbf{D}\right)\delta\mathbf{y} = \mathbf{E}\delta\mathbf{z} \tag{6.10}$$

Substituting Eqs. 6.6 and 6.8 into Eq. 6.10 gives:

$$\mathbf{EB}(\mathbf{I}-\mathbf{A})^{-1}\mathbf{D}^{-1}\mathbf{u} = \mathbf{v} \tag{6.11}$$

Equation 6.10 can alternatively be written as:

$$\mathbf{Gu} = \mathbf{v} \tag{6.12}$$

where

$$\mathbf{G} = \mathbf{EB}(\mathbf{I}-\mathbf{A})^{-1}\mathbf{D}^{-1} \tag{6.13}$$

where matrix **G** contains embedded information on the economic system's input–output linkages, as well as its connections with the environment that supplies its natural resource needs, and its baseline or normal level of production. Equation 6.12 will typically have excess degrees of freedom since a typical system will have more types of economic goods than natural resources.

Thus, if the fractional loss of natural resource supply (**v**) is given, a linear multiple-objective or vector optimization problem results:

$$\min \mathbf{u} \tag{6.14a}$$

subject to:

$$\mathbf{Gu} = \mathbf{v} \tag{6.14b}$$

Different approaches can be used to solve this problem. In the general case, the well-known ε-constraint method can be used to determine the Pareto frontier, which will contain all nondominated solutions. However, such an approach can lead to an infinite number of solutions that may not be sufficiently specific to be useful for decision support. Alternative methods for identifying unique solutions can also be used. For example, a simple additive weighting (SAW) approach can be used to convert the problem into a single-objective mathematical program:

$$\min \mathbf{w}^{\mathrm{T}}\mathbf{u} \tag{6.15a}$$

subject to:

$$\mathbf{Gu} = \mathbf{v} \tag{6.15b}$$

$$\sum_i \mathrm{w_i} = 1 \tag{6.15c}$$

where **w** is the vector of exogenously defined weighting factors associated with the economic sectors. As an additional formality, the weights can further be normalized to sum up to unity (Eq. 6.15c). However, there remains the issue of how the weights can be determined in the first place. One option is to set the weights based on the percentage contribution of the sectors to the GDP of the economic system. In this case, the model gives the solution that minimizes total reduction in GDP.

An alternative approach is to use fuzzy optimization as proposed by Zimmermann [10]. Instead of requiring a weight vector **w**, this formulation uses predefined limits for each element of **u**. Upper and lower bounds are used to define a linear *membership function* for each element in **u**. The lower limit corresponds to a lower magnitude of economic loss and is thus designated with a membership value of 1. Conversely, the upper limit gives the maximum tolerable threshold of economic loss and has a membership value of 0. Intermediate values have fractional membership values following a linear scale in the interval [0, 1]. The fuzzy optimization model is:

$$\max \lambda \tag{6.16a}$$

s.t.

$$\mathbf{Gu} = \mathbf{v} \tag{6.16b}$$

$$\left(u_j - u_j^{\max}\right)/\left(u_j^{\min} - u_j^{\max}\right) \geq \lambda \quad \forall j \tag{6.16c}$$

$$0 \leq \lambda \leq 1 \tag{6.16d}$$

where variable λ is the aggregate degree of membership (i.e., overall degree of satisfaction of fuzzy goals), $u_j^{\min}$ is the lower limit, and $u_j^{\max}$ is the upper limit of fraction loss of final demand of each sector *j*. To eliminate any computational issues arising from nonlinearity, this formulation uses max–min aggregation [10]. The aggregate degree of membership is equal to the lowest degree of membership in the individual membership functions of the sectors. This approach ensures that no sector receives a disproportionate penalty in the optimal solution; in other words, it may be said that max–min aggregation "protects the weakest link."

6.3 Case Study 6.1

This didactic case study is based on the two-sector illustrative example used by Tan et al. [9], which in turn was derived from Miller and Blair [8] with additional data in the form of natural resource inputs into the economic system. This is the same case study discussed in Chap. 2. The baseline or normal state of the system is given in Table 6.1 and is equivalent to Fig. 6.1.

From these data, it is possible to determine the matrices **A** and **B**, as shown in Table 6.2. The first two rows of data give the elements of matrix **A**, while the last row of data gives those of **B**.

Suppose that there is a disruption that causes 10% loss in natural resource supply. In a simplistic scenario, one possible solution is to have a proportionate 10% contraction of the entire economic system. However, such a solution is not necessarily optimal. If it is assumed that there is some flexibility in the reallocation of the natural resource (e.g., allocation of water for household, industry and agricultural use during a drought), total damage can be minimized by using the optimization model described in the previous section. The fuzzy limits for fractional loss of final demand in the two sectors must first be specified, as in Table 6.3. In practice, such limits can be exogenously defined in part by accounting for socioeconomic aspects (e.g., vulnerability, contribution to employment) that are not directly reflected within the model.

Using Eqs. 6.1–6.13, it is then possible to find matrix **G**. Since there is only one natural resource in this problem, the result is a row vector **G** = **g** = (0.1248 0.8752). The fuzzy optimization model can then be formulated in LINGO as:

Table 6.1 Economic and environmental flows in Case Study 6.1

	Sector 1	Sector 2	Final demand	Total output
Sector 1	150	500	350	1000
Sector 2	200	100	1700	2000
Resource	20	80	100[a]	n/a

[a]Total resource drawn by economy from environment

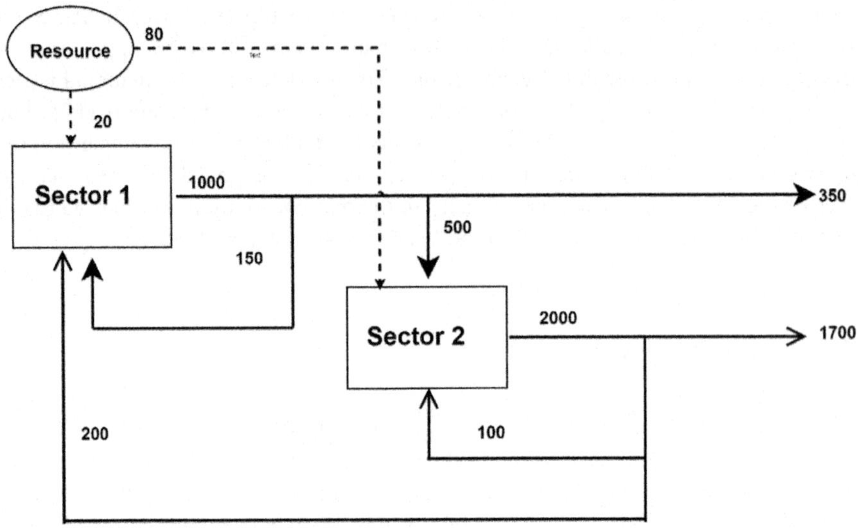

Fig. 6.1 Undisrupted economic system

Table 6.2 Coefficients of **A** and **B** in Case Study 6.1

	Sector 1	Sector 2
Sector 1	0.15	0.25
Sector 2	0.20	0.05
Resource	0.02	0.04

Table 6.3 Limits for fractional loss of final demand in Case Study 6.1

	u_j^{min}	u_j^{max}
Sector 1	0.05	0.10
Sector 2	0.04	0.12

```
max = lambda;                                  !Equation 6.16a;

0.1248*u1 + 0.8752*u2 = 0.10;                  !Equation 6.16b;
(u1 - 0.10) <= lambda*(0.05 - 0.10);           !Equation 6.16c;
(u2 - 0.12) <= lambda*(0.04 - 0.12);           !Equation 6.16c;
lambda < 1;                                    !Equation 6.16d;
```

When the LINGO model is solved, the following report is generated:

```
Global optimal solution found.
  Objective value:                              0.2295426
  Infeasibilities:                              0.000000
  Total solver iterations:                             2
  Elapsed runtime seconds:                          0.05

  Model Class:                                        LP

  Total variables:                    3
  Nonlinear variables:                0
  Integer variables:                  0

  Total constraints:                  5
  Nonlinear constraints:              0

  Total nonzeros:                     8
  Nonlinear nonzeros:                 0



                    Variable           Value        Reduced Cost
                      LAMBDA       0.2295426            0.000000
                          U1       0.8852287E-01        0.000000
                          U2       0.1016366            0.000000
```

The optimal value of λ is approximately 0.23, which implies that the fuzzy membership functions defined by the limits in Table 6.3 are 23% satisfied. At this state, the optimal fractional final demand losses of the two sectors are given in Table 6.4. While both sectors lose close to 10% of their respective final demands, there is a slightly higher level of fractional loss allocated to Sector 2 ($U2 = 10.16\%$) and slightly less to Sector 1 ($U1 = 8.85\%$). This result is because the data in Table 6.3 imply that Sector 2 can tolerate more fractional loss than Sector 1. For this solution, the GDP loss can be determined to be 203.7 units where 30.98 (8.852% * 350 = 30.98) is associated with the loss in Sector 1 while 172.72 (10.16% * 1700 = 172.72) is associated with the loss in Sector 2. The optimal network is illustrated in Fig. 6.2.

For comparison purposes, the same problem can be solved using the SAW approach (Eq. 6.15a–c). Based on the contribution to GDP under normal conditions, the weights corresponding to the sectors are 0.3333 and 0.6667, respectively. As previously discussed, using these weights should minimize economy-wide GDP loss.

The LINGO code for the model is:

```
min = 0.3333*u1 + 0.6667*u2;              !Equation 6.15a;

0.1248*u1 + 0.8752*u2 = 0.10;             !Equation 6.15b;
```

Table 6.4 Optimal values of fractional loss of final demand in Case Study 6.1

	Fuzzy model	SAW model
Sector 1	0.088	0
Sector 2	0.102	0.114

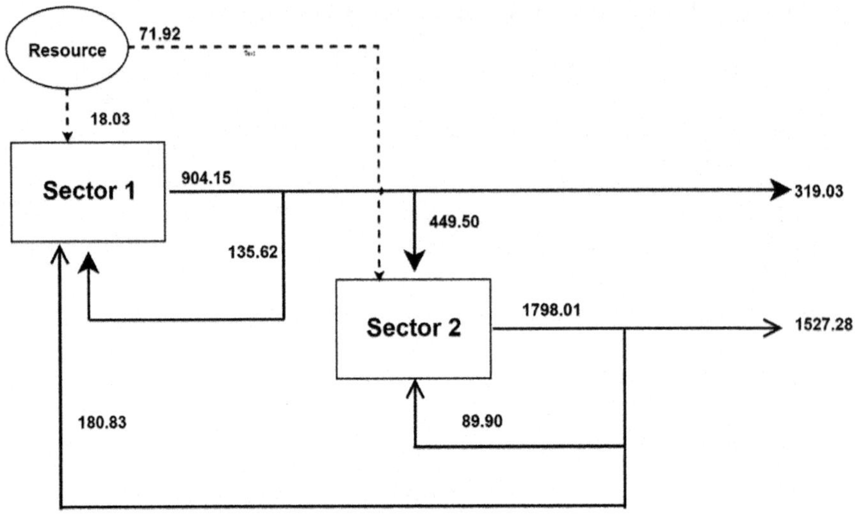

Fig. 6.2 Optimal network for disrupted system using fuzzy model

The corresponding solution report is:

```
Global optimal solution found.
  Objective value:                              0.7617687E-01
  Infeasibilities:                               0.000000
  Total solver iterations:                              0

  Model Class:                                         LP

  Total variables:                    2
  Nonlinear variables:                0
  Integer variables:                  0

  Total constraints:                  2
  Nonlinear constraints:              0

  Total nonzeros:                     4
  Nonlinear nonzeros:                 0

                    Variable           Value        Reduced Cost
                          U1        0.000000           0.2382313
                          U2       0.1142596            0.000000

                         Row    Slack or Surplus      Dual Price
                           1       0.7617687E-01       -1.000000
                           2        0.000000          -0.7617687
```

The GDP loss in this case can be calculated to be 194.31 units (11.43% * 1700 = 194.31), which is much lower than the 203.7 units that results from the fuzzy solution. The resulting network is illustrated in Fig. 6.3. However, this gain comes at the expense of a less equitable distribution of losses. In the final

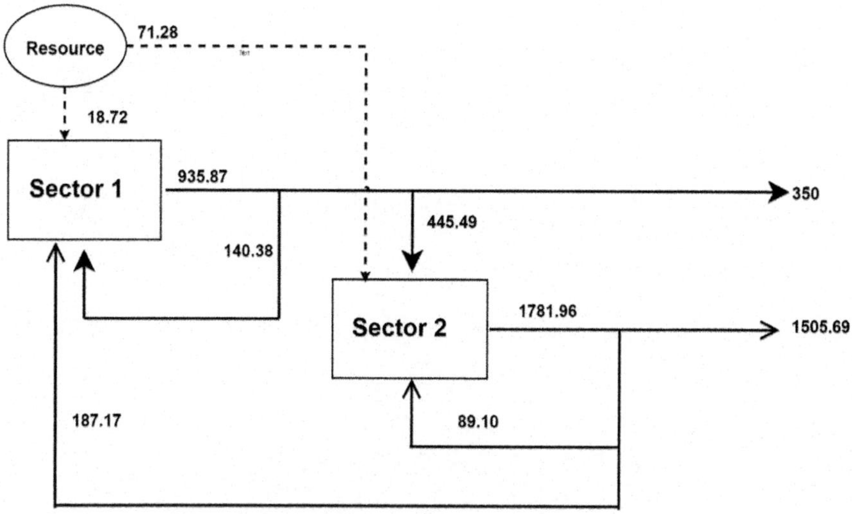

Fig. 6.3 Optimal network of disrupted system using SAW model

column of Table 6.4, it can be seen that all the final demand loss occurs in Sector 2. In other words, as much Sector 2 output is diverted from final demand as is needed to ensure that the intermediate demand of Sector 1 is first met. Because of the compensatory nature of the SAW formulation, it can potentially result in heavy losses to be incurred disproportionately by one sector, for as long as such concentrated losses are offset elsewhere in the economic system to give the best aggregate objective function value. The fuzzy model on the other hand ensures that the losses are distributed more equitably.

6.4 Sample Case Study 6.2

This case study demonstrates the use of the model for allocating land resources that are degraded as a result of climate change. We consider the case of Malaysia, where the oil palm sector accounts for about 36% of total agricultural output, according to the 2005 input–output data from the Department of Statistics Malaysia [2]. Oil palm plantations also account for 4.92 million hectares out of the 7.87 million hectares of total agricultural land in the country according to the Malaysian Palm Oil Board [7]. In this case study, a hypothetical scenario of 10% loss of land resource is considered. The input–output data has been aggregated into 12 sectors following the procedure outlined in Miller and Blair [8]. Table 6.5 shows the coefficients of matrices **A** and **B** for this case study; the final row corresponds to agricultural land resource and comprises **B**. The final demand vector is indicated in the last column of Table 6.5. It is then possible to obtain matrix **G** using Eq. 6.13 with the resulting matrix shown below. Also, the upper and lower limits for fractional final demand degradation are given in Table 6.6.

Table 6.5 Coefficients of **A** and **B** in Case Study 6.2

	Agriculture, fishery, and forestry excluding palm oil	Palm oil	Mining and quarrying	Manufacturing	Electricity, gas and water	Construction	Trade	Transportation, communication, and storage	Finance	Real estate and ownership of dwellings	Private services	Government services	Final demand (in thousand Ringgit)
Agriculture, fishery, and forestry excluding palm oil	0.114	0.013	0.002	0.019	0.000	0.001	0.014	0.000	0.001	0.000	0.000	0.003	14,510,227.55
Palm oil	0.001	0.091	0.001	0.020	0.000	0.000	0.001	0.000	0.000	0.000	0.000	0.001	1,625,892.69
Mining and quarrying	0.003	0.002	0.005	0.042	0.018	0.032	0.004	0.001	0.002	0.000	0.001	0.003	53,901,753.33
Manufacturing	0.144	0.108	0.100	0.324	0.223	0.305	0.100	0.106	0.030	0.028	0.063	0.101	530,734,250.54
Electricity, gas and water	0.004	0.010	0.002	0.009	0.186	0.004	0.008	0.010	0.007	0.013	0.031	0.020	13,054,117.50
Construction	0.000	0.000	0.007	0.000	0.043	0.001	0.004	0.022	0.000	0.010	0.001	0.086	35,400,608.90
Trade	0.024	0.025	0.044	0.047	0.010	0.027	0.043	0.012	0.025	0.139	0.011	0.037	25,631,176.53
Transportation, communication, and storage	0.014	0.015	0.013	0.021	0.013	0.043	0.044	0.262	0.070	0.017	0.047	0.051	47,413,485.55
Finance	0.003	0.031	0.000	0.009	0.004	0.032	0.022	0.108	0.300	0.021	0.023	0.005	23,460,682.99
Real estate and ownership of dwellings	0.000	0.000	0.001	0.000	0.001	0.007	0.014	0.023	0.013	0.081	0.044	0.030	19,766,906.97
Private services	0.001	0.006	0.006	0.003	0.015	0.026	0.039	0.023	0.029	0.017	0.188	0.037	38,796,551.42
Government services	0.000	0.000	0.001	0.000	0.004	0.005	0.004	0.015	0.004	0.007	0.009	0.064	70,027,405.50
Agricultural land	0.077	0.226	0.000	0.000	0.000	0.000	0.000	0.000	0.000	0.000	0.000	0.000	7,920,000 (Ha)

Table 6.6 Limits for fractional loss of final demand in Case Study 6.2

	$u_j^{\min}$	$u_j^{\max}$
Agriculture, fishery, and forestry excluding palm oil	0	0.2
Palm oil	0	0.3
Mining and quarrying	0	0.1
Manufacturing	0	0.1
Electricity, gas, and water	0	0.1
Construction	0	0.1
Trade	0	0.1
Transportation, communication, and storage	0	0.1
Finance	0	0.1
Real estate and ownership of dwellings	0	0.1
Private services	0	0.1
Government services	0	0.1

$$\mathbf{G} = [0.1642 \quad 0.0519 \quad 0.0105 \quad 0.6973 \quad 0.0054 \quad 0.0165 \quad 0.0091 \quad 0.0123 \quad 0.0032 \quad 0.0024 \quad 0.0061 \quad 0.0207]$$

The corresponding set-based LINGO code is given below:

```
SETS:
SECTORS:  U,   UL, UU;
RESOURCE: V;
BMATRIX(RESOURCE, SECTORS): G;

ENDSETS

DATA:
N = 12;
SECTORS = 1..N;
RESOURCE = LAND;

G =
0.16425      0.05191      0.01045      0.69729      0.0054       0.01654
0.00908      0.01227      0.003192     0.002436     0.006149     0.02074;

V = 0.10;
UL =  0      0      0      0      0      0      0      0      0      0      0      0;
UU =  0.20   0.30   0.1    0.1    0.1    0.1    0.1    0.1    0.1    0.1    0.1
      0.1;
ENDDATA

MAX = LAMBDA;              !Equation 6.16a;
@FOR(RESOURCE(I): @SUM(SECTORS(J): G(I,J)*U(J)) = V(I)); !Equation 6.16b;
@FOR(SECTORS(J): (U(J) - UU(J))<= LAMBDA*(UL(J) - UU(J))); !Equation 6.16c;
LAMBDA<1;    !Equation 6.16d;
```

Solving this model gives the optimal level of fractional final demand losses of the 12 sectors, as shown in the first data column of Table 6.7 and indicated in the solution report.

Table 6.7 Optimal values of fractional loss of final demand in Case Study 6.2

	Fuzzy model
Agriculture, fishery, and forestry excluding palm oil	0.158
Palm oil	0.236
Mining and quarrying	0.079
Manufacturing	0.079
Electricity, gas and water	0.079
Construction	0.079
Trade	0.079
Transportation, communication, and storage	0.079
Finance	0.079
Real estate and ownership of dwellings	0.079
Private services	0.079
Government services	0.079

```
Global optimal solution found.
Objective value:                              0.2112177
Infeasibilities:                               0.000000
Total solver iterations:                             13
Elapsed runtime seconds:                           0.13

  Model Class:                                          LP

  Total variables:                      13
  Nonlinear variables:                   0
  Integer variables:                     0

  Total constraints:                    15
  Nonlinear constraints:                 0

  Total nonzeros:                       38
  Nonlinear nonzeros:                    0


                        Variable           Value        Reduced Cost
                               N        12.00000            0.000000
                          LAMBDA       0.2112177            0.000000
                           U( 1)       0.1577565            0.000000
                           U( 2)       0.2366347            0.000000
                           U( 3)       0.7887823E-01        0.000000
                           U( 4)       0.7887823E-01        0.000000
                           U( 5)       0.7887823E-01        0.000000
                           U( 6)       0.7887823E-01        0.000000
                           U( 7)       0.7887823E-01        0.000000
                           U( 8)       0.7887823E-01        0.000000
                           U( 9)       0.7887823E-01        0.000000
                          U( 10)       0.7887823E-01        0.000000
                          U( 11)       0.7887823E-01        0.000000
                          U( 12)       0.7887823E-01        0.000000
```

Note that final demand loss for nonpalm agriculture (Sector 1) is 15.8%, while the corresponding loss for oil palm (Sector 2) is 23.6%. The loss for all other sectors in the economy is 7.9%, leading to an overall GDP loss of 8.0% relative to the baseline of Ringgit 874 billion. This solution corresponds to an optimal value of

$\lambda = 0.21$. It can be seen that the losses are equitably distributed, since the model structure ensures that their values fall within the a priori bounds given in Table 6.6.

Alternatively, a SAW-based model can be formulated in LINGO to minimize GDP loss instead. Using equal weights for all sectors, the model can be formulated as follows:

```
SETS:
SECTORS:  U,  W;
RESOURCE: V;
BMATRIX(RESOURCE, SECTORS): G;

ENDSETS

DATA:
N = 12;
SECTORS = 1..N;
RESOURCE = LAND;

G =
0.16425     0.05191     0.01045     0.69729     0.0054      0.01654
0.00908     0.01227     0.003192    0.002436    0.006149    0.02074;
V = 0.10;
ENDDATA

MIN = @SUM(SECTORS(J): W(J)*U(J));
@FOR(RESOURCE(I): @SUM(SECTORS(J): G(I,J)*U(J)) = V(I));
@FOR(SECTORS(J): W(J) = 1/N);
```

Solving this model generates the following solution report:

```
Global optimal solution found.
  Objective value:                              0.1195103E-01
  Infeasibilities:                               0.000000
  Total solver iterations:                              0
  Elapsed runtime seconds:                           0.14

  Model Class:                                         LP

  Total variables:                 12
  Nonlinear variables:              0
  Integer variables:                0

  Total constraints:                2
  Nonlinear constraints:            0

  Total nonzeros:                  24
  Nonlinear nonzeros:               0


                    Variable          Value        Reduced Cost
                           N       12.00000            0.000000
                       U( 1)       0.000000        0.6370377E-01
                       U( 2)       0.000000        0.7712955E-01
                       U( 3)       0.000000        0.8208445E-01
                       U( 4)      0.1434124            0.000000
                       U( 5)       0.000000        0.8268798E-01
                       U( 6)       0.000000        0.8135663E-01
                       U( 7)       0.000000        0.8224818E-01
                       U( 8)       0.000000        0.8186694E-01
                       U( 9)       0.000000        0.8295186E-01
                      U( 10)       0.000000        0.8304221E-01
                      U( 11)       0.000000        0.8259846E-01
                      U( 12)       0.000000        0.8085469E-01
```

Table 6.8 Sector weights and optimal values of fractional loss of final demand in Case Study 6.2

	Weight	SAW model
Agriculture, fishery, and forestry excluding palm oil	0.0833	0
Palm oil	0.0833	0
Mining and quarrying	0.0833	0
Manufacturing	0.0833	0.143
Electricity, gas, and water	0.0833	0
Construction	0.0833	0
Trade	0.0833	0
Transportation, communication, and storage	0.0833	0
Finance	0.0833	0
Real estate and ownership of dwellings	0.0833	0
Private services	0.0833	0
Government services	0.0833	0

As in the previous case study, minimum GDP loss occurs when loss of final demand is concentrated in the disrupted sector. This approach may be optimal from the global perspective, but fails to account for the need for equity in practical decision-making (Table 6.8).

6.5 Conclusion

Input–output models of supply chain networks can be used to optimize responses to loss of natural resource inputs by providing insights on how to best allocated production capacity during the crisis. This capability can make such models useful for ensuring sustainable and resilient systems. In particular, optimization models based on the input–output framework can be used for this purpose. These problems often lead to multiple-objective optimization scenarios that can be solved using different aggregation techniques. In this chapter, a model using max–min aggregation within a fuzzy linear programming model was used, with two case studies illustrating its capabilities. Alternative aggregation approaches can also be used with minor modifications in the model.

References

1. Chen K (1973) Input-output economic analysis of environmental impact. IEEE Trans Syst, Man Cybern 3:539–547
2. Department of Statistics Malaysia. (http://www.statistics.gov.my/portal/index.php?option=com_content&view=article&id=1242&Itemid=111&lang=en. Accessed 27 Apr 2013)
3. Jiang P, Haimes YY (2004) Risk management for Leontief-based interdependent systems. Risk Anal 24:1215–1229

4. Kananen I, Korhonen P, Wallenius J, Wallenius H (1990) Multiple objective analysis of input–output models for emergency management. Oper Res 38:193–201
5. Khanna V, Bakshi BR (2009) Modeling the risks to complex industrial networks due to loss of natural capital. In: IEEE international symposium on sustainable systems and technology, ISSST '09 in cooperation with 2009 IEEE international symposium on technology and society, ISTAS, art. no. 515677
6. MacKenzie CA, Santos JR, Barker K (2012) Measuring changes in international production from a disruption: case study of the Japanese earthquake and tsunami. Int J Prod Econ 138:293–302
7. Malaysian Palm Oil Board. (http://www.palmoilworld.org/about_malaysian-industry.html. Accessed 27 Apr 2013)
8. Miller RE, Blair PD (2009) Input–output analysis: foundations and extensions, 2nd edn. University Press, Cambridge, UK
9. Tan RR, Aviso KB, Promentilla MAB, Yu KDS, Santos JR (2014) Development of a fuzzy linear programming model for allocation of inoperability in economic sectors due to loss of natural resource inputs. DLSU Bus Econ Rev 24:1–12
10. Zimmermann H-J (1978) Fuzzy programming and linear programming with several objective functions. Fuzzy Sets Syst 1:45–55

Chapter 7
Life Cycle Assessment Models

Abstract Life cycle assessment (LCA), a comprehensive framework for analyzing the environmental impacts of products and systems on a cradle-to-grave paradigm, is a methodology closely related to environmentally extended input–output analysis. In this chapter, a discussion of the mathematical principles of LCA is given. Two illustrative case studies are also solved, and the corresponding LINGO codes are provided.

Keywords Life cycle assessment · Goal and scope definition · Inventory analysis Impact assessment · Interpretation

7.1 Introduction

Life cycle assessment (LCA) is a systematic methodology for the analysis of the environmental impacts of product systems from a "cradle-to-grave" perspective. A product system can be broadly defined as the network of processes or activities needed to deliver a product (or service) to an end user. Based on this framework, environmental impacts embodied in input streams into a given process add to the impacts that occur directly in the process itself, such that a complete environmental profile of a product system can be determined. LCA methodology in its modern form was developed as a result of a series of technical workshops by the Society of Environmental Toxicology and Chemistry (SETAC) in the early 1990s, based on forerunners such as resource and environmental profile analysis (REPA) and net energy analysis. Fava et al. [1] give a good historical account of the early days of the development of this methodology. Subsequent to these early efforts came the emergence of the ISO 14040-43 standards in the late 1990s, and more recently the publication of the ISO 14040 and 14044 Standards in 2006. An account of the development of these standards, as well as recent extensions of LCA such as carbon footprint and water footprint, is given in a chapter by Finkbeiner [2]. This chapter focuses on the computational aspects of LCA, its similarities to I–O modeling, and the implementation of LCA models using LINGO.

R. R. Tan et al., *Input–Output Models for Sustainable Industrial Systems*,
Lecture Notes in Management and Industrial Engineering,
https://doi.org/10.1007/978-981-13-1873-3_7

7.2 Life Cycle Assessment

LCA utilizes some key underlying principles to ensure a rational evaluation of the environmental impacts of a product system. Firstly, as the term suggests, LCA extends the boundaries of a product system using the aforementioned "cradle-to-grave" perspective. Conceptually, each product system should be analyzed by tracing all upstream process chains to their ultimate sources (i.e., extraction of resources from the natural environment) and likewise by tracing all downstream process chains to their final destinations (i.e., release of emissions and wastes into the environment). Thus, the analysis in principle encompasses the entire life cycle of the product. In practice, the analysis of such comprehensive product systems may prove to be computationally cumbersome or intractable, and analysts who apply LCA apply simplifying assumptions by setting system boundaries such that the resulting life cycles are of reasonable size.

Analysis is also done on the basis of a predefined unit of output of a product system, known as the *functional unit*. In the most general case, the functional unit represents a specific unit of service or value delivered (e.g., person-kilometers in the case of mass transport systems), although physical quantities (e.g., measured in mass or energy units) are also used in some cases. The functional unit allows proper benchmarking in cases where LCA results are used for comparison of alternatives. In addition, LCA is intended to provide a multi-dimensional scorecard for the product system being analyzed, by evaluating its performance with respect to a predefined range of environmental impacts such as global warming, stratospheric ozone depletion, eutrophication. In practice, such environmental impact scores can be used for conducting a thorough evaluation, even though the element of subjectivity is impossible to eliminate given the current state of the art of LCA methodology.

LCA thus naturally necessitates a quantitative approach, analogous to financial bookkeeping, to ensure that environmental impacts are properly accounted for. The four components of LCA as outlined in the ISO 14040 and 14044 standards are:

- **Goal and scope definition**—Identification of purpose, system boundaries, functional unit, technological assumptions, the natural resources, pollutants and environmental impacts of interest, data sources, and other relevant assumptions
- **Life cycle inventory analysis** (LCI)—Estimation of flows of natural resources into and pollutants from the product system per functional unit
- **Life cycle impact assessment** (LCIA)—Estimation of environmental impacts of the product system per functional unit
- **Interpretation**—Use of information derived from previous steps to address the purpose of the LCA and to determine whether or not the results are sufficiently conclusive given the errors and uncertainties that occur in the analysis.

The LCI component is of particular interest here due to its close mathematical similarity to I–O analysis [4, 6]. The generic process-based LCI model is:

$$\mathbf{Ps} = \mathbf{f} \tag{7.1}$$

where **P** is the $m \times n$ process matrix, **s** is the scaling vector with n elements, and **f** is the functional unit vector with m elements. Each row of the process matrix corresponds to a stream of an economic commodity (i.e., a product or an intermediate), while each column corresponds to a process or activity at an appropriate level of aggregation. The numerical values of the coefficients of **P** assume positive and negative values for output and input streams, respectively. Each column vector $\mathbf{p}_j$ in matrix **P** represents a process, and the relative values of the elements in the column vector represent the relative proportions of stream flowrates (as determined from process stoichiometry, thermodynamics, etc.). Thus, the values in matrix **P** represent the state of technology of the product system. On the other hand, vector **f** is exogenously defined based on the selected functional unit for the LCA problem. Generally, only the row corresponding to the final product will assume a nonzero value, as it is assumed that the other intermediate products are consumed entirely within the product system. In this sense, the LCI model isolates the final product. It is then necessary to solve for **s** given **P** and **f** using Eq. 7.1. Depending on the relative values of m and n, Eq. 7.1 can have negative, zero, or positive degrees of freedom. Negative degrees of freedom ($m > n$) indicate multi-functionality (i.e., the presence of processes that generate more than one product), which can result in difficulties in attribution of environmental impacts to a specific main product. Such cases can be dealt with using approaches such as allocation, displacement, or system expansion, which mathematically entail modification of matrix **P** so that the final condition is $m = n$. Details of these procedures are discussed in Heijungs and Suh [4]. Positive degrees of freedom ($m < n$) indicate the presence of alternative processes and provide opportunities for identifying an optimal configuration for the product system. A brief discussion of this scenario is given in a chapter by Tan et al. [6].

For systems with zero degrees of freedom ($m = n$), and provided **P** is nonsingular, the solution can be determined by matrix inversion as in I–O models:

$$\mathbf{s} = \mathbf{P}^{-1}\mathbf{f} \tag{7.2}$$

This inversion step is analogous to determining sector output from final demand and the Leontief inverse in I–O models. Next, the net flow of environmental streams (resources and pollutants) from the product system can be determined as:

$$\mathbf{Qs} = \mathbf{g} \tag{7.3}$$

where **Q** is the $p \times n$ process environmental matrix and **g** is the product system inventory vector with p elements. The values in **Q** follow the same input and output conventions as used for **P**. Each column vector $\mathbf{q}_j$ in **Q** is an extension of the corresponding column vector $\mathbf{p}_j$ in **P** and lists the flowrates of resources and pollutants relative to each unit of process output. Matrices **P** and **Q** are thus linked by a common scaling vector **s**. It thus becomes possible to calculate from mass and energy balance principles the quantity of natural resources consumed and pollutants generated (**g**) for a given functional unit (**f**).

Equations 7.2 and 7.3 can be combined to give the generic LCI model in simplified form:

$$\mathbf{QP}^{-1}\mathbf{f} = \mathbf{g} \tag{7.4}$$

In LCA, it is then necessary to proceed to an LCIA step to estimate potential environmental impacts arising from the LCI results. In particular, the links can be quantified via characterization:

$$\mathbf{Rg} = \mathbf{h} \tag{7.5}$$

where **R** is the $q \times p$ characterization matrix and **h** is the impact vector with q elements. The purpose of **R** is to map flows of natural resources and pollutants in vector **g** to corresponding impact categories in **h**, and the numerical values of its elements signify the magnitude of contribution of each unit of flow to a particular impact. The impact scores are usually sufficient for purposes of interpretation. In some cases, there are subsequent steps to normalize the impact scores or to aggregate them into a single overall metric. However, such steps are beyond the scope of this chapter. This generic LCA model is illustrated in the following sections using case studies from the literature, with emphasis on implementation using LINGO.

7.3 Case Study 7.1

This example is based on a simple example by Heijungs [3] and is limited only to LCI. The case study uses fictitious data and is intended only to illustrate the computational principles discussed in the previous section. The product system consists of four major processes, namely:

- Production of electricity
- Production of aluminum
- Production of aluminum foil
- Usage of aluminum foil to wrap sandwiches.

The economic streams in the product system are:

- Electricity (measured in MJ)
- Aluminum (measured in kg)
- Aluminum foil (measured in kg)
- Sandwich packages (measured in number of pieces).

Finally, the environmental streams are:

- Bauxite (measured in kg)
- Crude oil (measured in kg)
- CO_2 (measured in kg)
- Solid waste (measured in kg).

For this product system, the elements of **P** are given in Table 7.1, while the elements of **Q** are given in Table 7.2. The schematic diagram of the process units is illustrated in Fig. 7.1.

Given these process data, we wish to determine the amounts of bauxite and crude oil consumed, and CO_2 and solid waste generated, for a functional unit of ten sandwich packages. Thus, the problem is to compute the inventory vector **g**, given the functional unit vector $\mathbf{f} = (0 \quad 0 \quad 0 \quad 10)^T$. The LCI model can be determined by coding Eqs. 7.1 and 7.3 in LINGO as follows:

```
!Equation 7.1;
1*s1 - 50*s2 - 1*s3 + 0*s4 = 0;
-0.01*s1 + 1*s2 - 1*s3 + 0*s4 = 0;
0*s1 + 0*s2 + 1*s3 - 1*s4 = 0;
0*s1 + 0*s2 + 0*s3 + 1*s4 = 10;
```

Table 7.1 Elements of **P** for Case Study 7.1

	Process			
	Production of electricity	Production of aluminum	Production of aluminum foil	Usage of aluminum foil
Electricity (MJ)	1	−50	−1	0
Aluminum (kg)	−0.01	1	−1	0
Aluminum foil (kg)	0	0	1	−1
Sandwich packages (pieces)	0	0	0	100

Table 7.2 Elements of **Q** for Case Study 7.1

	Process			
	Production of electricity	Production of aluminum	Production of aluminum foil	Usage of aluminum foil
Bauxite (kg)	0	−5	0	0
Crude oil (kg)	−0.5	0	0	0
CO_2 (kg)	3	0	0	0
Solid waste (kg)	2	10	0	1

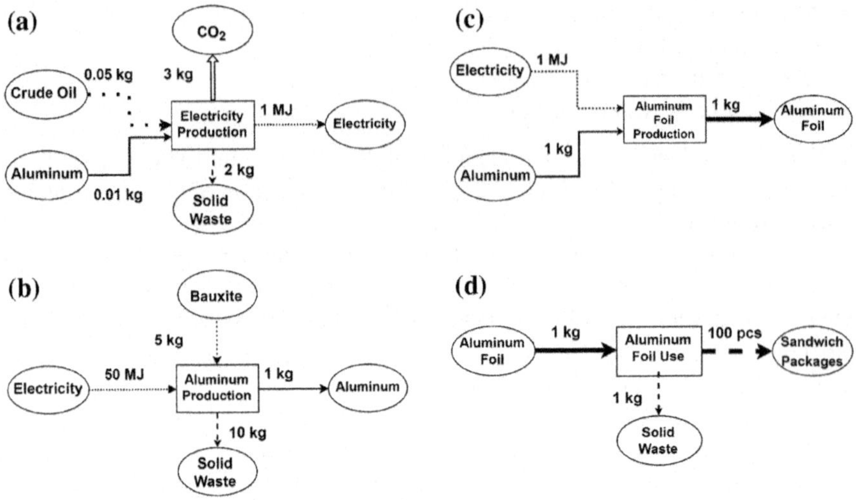

Fig. 7.1 Schematic diagram of **a** electricity production, **b** aluminum production, **c** aluminum foil production, and **d** aluminum foil use

```
!Equation 7.3;
0*s1 - 5*s2 + 0*s3 + 0*s4 = g1;
-0.5*s1 + 0*s2 + 0*s3 + 0*s4 = g2;
3*s1 + 0*s2 + 0*s3 + 0*s4 = g3;
2*s1 + 10*s2 + 0*s3 + 1*s4 = g4;
```

```
!Additional conditions;
@free(g1);
@free(g2);
```

Note that the LINGO @free command is used in anticipation of negative values for resource flowrates g_1 and g_2. Then, LINGO can be prompted to find the solution to this system of equations. The resulting solution report is:

```
Feasible solution found.
   Infeasibilities:                          0.000000
   Total solver iterations:                         0

   Model Class:                                    LP

   Total variables:                                 8
   Nonlinear variables:                             0
   Integer variables:                               0

   Total constraints:                               9
   Nonlinear constraints:                           0

   Total nonzeros:                                 19
   Nonlinear nonzeros:                              0


Variable           Value
      S1        10.20000
      S2       0.2020000
      S3       0.1000000
      S4       0.1000000
      G1       -1.010000
      G2       -5.100000
      G3        30.60000
      G4        22.52000
```

It can be seen from this solution report that the scaling vector $\mathbf{s} = (\,10.2 \quad 0.202 \quad 0.1 \quad 0.1\,)^{\mathrm{T}}$, and that the inventory vector $\mathbf{g} = (\,-1.01 \quad -5.1 \quad 30.6 \quad 22.52\,)^{\mathrm{T}}$, which matches the original result of Heijungs [3]. Thus, for the given product system, the functional unit of ten sandwich packages requires the extraction of 1.01 kg of bauxite and 5.1 kg of crude oil, while also generating 30.6 kg of CO_2 and 22.52 kg of solid waste. All of these resource flows are distributed throughout the product system. The balanced flows are shown in Table 7.3 and illustrated in Fig. 7.2.

Table 7.3 Balanced process flows for Case Study 7.1

	Process			
	Production of electricity	Production of aluminum	Production of aluminum foil	Usage of aluminum foil
Electricity (MJ)	10.2	−10.1	−0.1	0
Aluminum (kg)	−0.102	0.202	−0.1	0
Aluminum foil (kg)	0	0	0.1	−0.1
Sandwich packages (pieces)	0	0	0	10
Bauxite (kg)	0	−1.01	0	0
Crude oil (kg)	−5.1	0	0	0
CO_2 (kg)	30.6	0	0	0
Solid waste (kg)	20.4	2.02	0	0.1

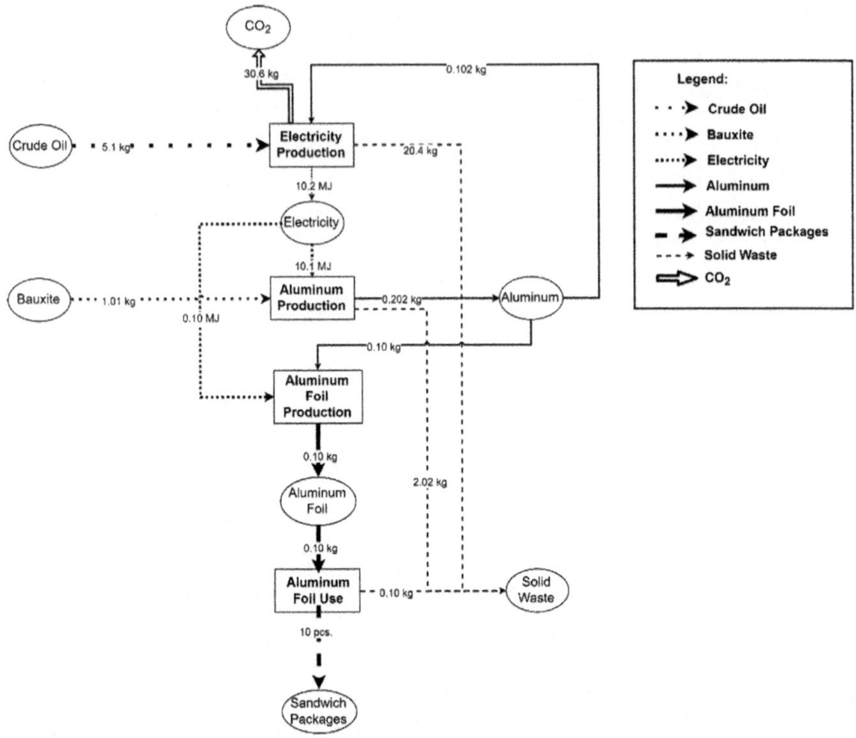

Fig. 7.2 Balanced system of Case Study 7.1

Once the LCI model is established, specific process technology changes can be simulated by changing the appropriate coefficients in matrices **P** and **Q**. For example, two changes can be analyzed here:

- **Process Change 1**. Improving the efficiency of the aluminum production process so as to reduce electricity consumption per unit of aluminum by 4%. This change can be implemented by reducing the absolute value of $a_{1,2}$ by 4%, from −50 MJ in Table 7.1 to −48 MJ.
- **Process Change 2**. Reducing the thickness of the aluminum foil by 20%. This change allows more sandwich packages to be made per unit mass of aluminum foil, allowing $a_{4,4}$ to be increased from 100 pieces in Table 7.1 to 125 pieces. However, rolling the foil much thinner also increases electricity consumption in the production process by, say, 15%. This effect can be incorporated by simultaneously changing $a_{1,3}$ from −1 MJ in Table 7.1 to −1.15 MJ.

The solutions to these two alternative scenarios can be determined by making the appropriate changes in the LINGO model and solving the resulting modified programs. Figure 7.3 illustrates the flow of streams resulting from Process Change 1, while Fig. 7.4 shows the result when Process Change 2 is implemented. Table 7.4

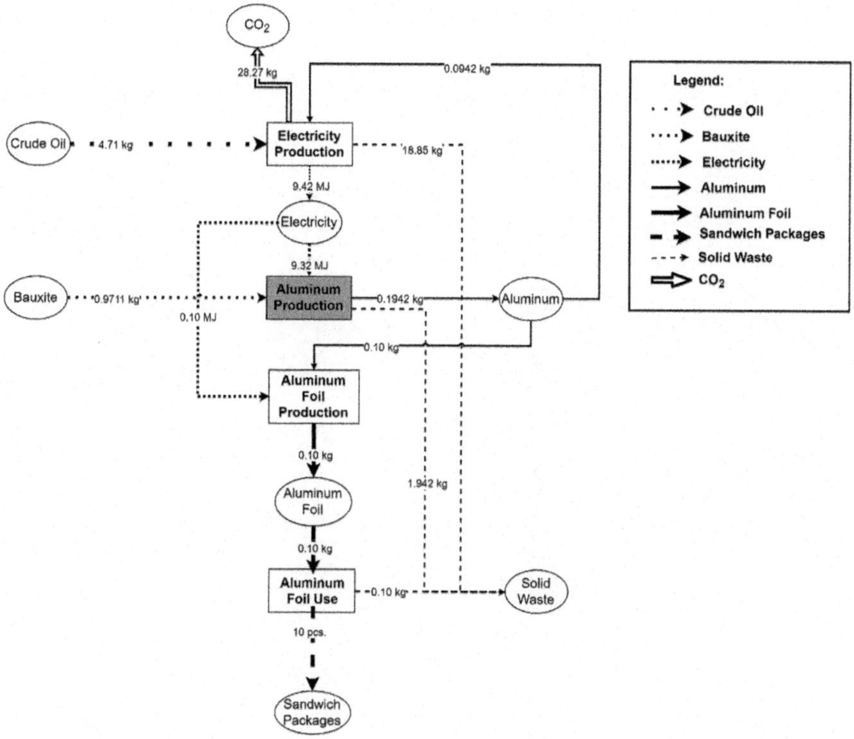

Fig. 7.3 Balanced system when Process Change 1 is implemented

shows the effects of these process changes compared to the baseline results of the case study. Note that, for these the first process change involving efficiency improvement in aluminum production, even a 4% reduction in electricity consumption results in 3.9% reduction in bauxite consumption, 7.6% reduction in both crude oil consumption and CO_2 emissions, and 7.2% reduction in solid waste generation. It can be seen that the percentage changes in three of the four elements of the inventory vector **g** are much larger than the relative change in the energy consumption. This amplification effect is the result of the feedback loop between the production processes for electricity and aluminum and is discussed in more detail by Heijungs and Suh [4]. The reductions that result from the second process change are even more pronounced, at 19.8% for each of the elements of the inventory vector **g**.

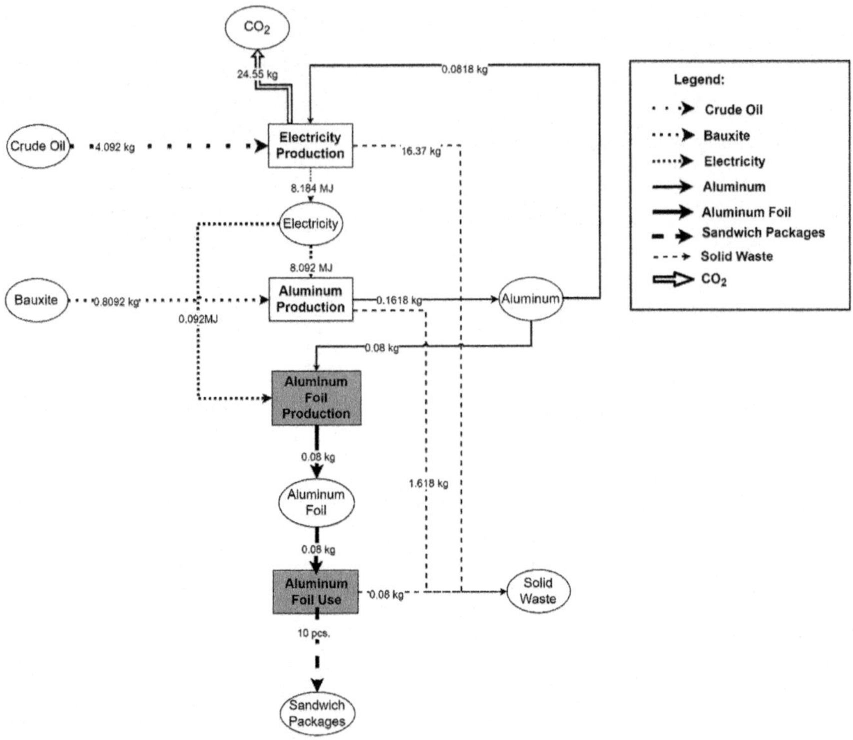

Fig. 7.4 Balanced system when Process Change 2 is implemented

Table 7.4 Effects of process changes in Case Study 7.1

	Scenario		
	Baseline	Process change 1	Process change 2
Bauxite (kg)	−1.01	−0.97	−0.81
Crude oil (kg)	−5.1	−4.71	−4.09
CO_2 (kg)	30.6	28.3	24.55
Solid waste (kg)	22.52	20.89	18.06

7.4 Case Study 7.2

This next example is based on the incandescent bulb case study of Heijungs [5] and covers both LCI and LCIA. The figures used here are hypothetical and are only for illustrative purposes. This product system consists of seven major processes, namely:

- P1: Production of electricity
- P2: Production of glass
- P3: Production of copper
- P4: Production of fuel oil
- P5: Production of bulbs
- P6: Usage of bulbs
- P7: Disposal of bulbs.

The economic streams in the product system are:

- S1: Electricity (measured in MJ)
- S2: Glass (measured in kg)
- S3: Copper (measured in kg)
- S4: Fuel oil (measured in kg)
- S5: New bulbs (measured in number of pieces)
- S6: Light (measured in hours of use of incandescent bulbs)
- S7: Spent bulbs (measured in number of pieces).

The environmental streams are:

- G1: Sand (measured in kg)
- G2: Copper ore (measured in kg)
- G3: Crude oil (measured in kg)
- G4: CO_2 (measured in kg emitted to air)
- G5: SO_2 (measured in kg emitted to air)
- G6: Copper (measured in kg released to soil).

Finally, the environmental impacts of interest are:

- H1: Resource depletion (measured in resource depletion units)
- H2: Eco-toxicity (measured in m^2 of ecosystem area)
- H3: Global warming (measured in kg CO_2 eq.)
- H4: Acidification (measured in kg CO_2 eq.).

For this product system, the elements of **P** are given in Table 7.5, while the elements of **Q** are given in Table 7.6.

For the LCIA phase, the characterization matrix **R** is given in Table 7.7. These characterization factors give the potential contribution of a unit of each resource or pollutant to a particular impact category. For example, every kg of sand consumed results in an impact of 0.1 Resource Depletion Units (RDUs), which scales the consumption of the resource to its abundance in the natural environment. Thus, copper ore and crude oil, which are much scarcer resources than sand, have correspondingly larger characterization factors.

Given these data, we now wish to find the resources consumed, pollutants generated, and environmental impacts arising from 1,000 h of use of an incandescent bulb. In other words, the problem is to compute the inventory vector **g** and impact vector **h** for a functional unit vector $\mathbf{f} = (0 \quad 0 \quad 0 \quad 0 \quad 1{,}000 \quad 0)^{\mathrm{T}}$. Coding each equation explicitly as in the previous case study, the LCA model can

Table 7.5 Elements of **P** for Case Study 7.2

	Process						
	Production of electricity	Production of glass	Production of copper	Production of fuel oil	Production of bulb	Use of bulb	Disposal of bulb
Electricity (MJ)	10^6	−100	−10,000	0	−1,000	−10,000	0
Glass (kg)	0	1,000	0	0	−10	0	0
Copper (kg)	0	0	100	0	−5	0	0
Fuel oil (kg)	−500	0	0	1,000	0	0	0
New bulb (pieces)	0	0	0	0	1,000	−1	0
Light (h)	0	0	0	0	0	5,000	0
Spent bulb (pieces)	0	0	0	0	0	1	−100

Table 7.6 Elements of **Q** for Case Study 7.2

	Process						
	Production of electricity	Production of glass	Production of copper	Production of fuel oil	Production of bulb	Use of bulb	Disposal of bulb
Sand (kg)	0	−1,000	0	0	0	0	0
Copper ore (kg)	0	0	−1,000	0	0	0	0
Crude oil (kg)	0	0	0	−1,200	0	0	0
CO_2 (kg)	1,000	0	0	0	0	0	100
SO_2 (kg)	100	0	0	0	0	0	0
Copper to soil (kg)	0	0	0	0	0	0	2

Table 7.7 Elements of **R** for Case Study 7.2

	Sand	Copper ore	Crude oil	CO_2	SO_2	Copper to soil
Resource depletion (RDU/kg)	0.1	10,000	250	0	0	0
Eco-toxicity (m^2/kg)	0	0	0	0	20	10,000
Global warming (kg CO_2 eq./kg)	0	0	0	1	0	0
Acidification (kg SO_2 eq./kg)	0	0	0	0	1	0

be determined by coding Eqs. 7.1, 7.3, and 7.5 in LINGO (using a set-based approach) as follows:

```
SETS:
PROCESS: S ;
STREAM: F;
ENVT: G;
IMPACT: H;
PMATRIX(STREAM, PROCESS): P;
QMATRIX(ENVT, PROCESS): Q;
RMATRIX(IMPACT, ENVT): R;
ENDSETS

DATA:

PROCESS =   P1     P2     P3     P4     P5     P6     P7;
STREAM =    S1     S2     S3     S4     S5     S6     S7;
ENVT =      G1     G2     G3     G4     G5     G6;
IMPACT =    H1     H2     H3     H4;

P =     1E6    -100   -10000        0      -1000  -10000       0
        0      1000   0             0      -10    0            0
        0      0      100           0      -5     0            0
        -500   0      0             1000   0      0            0
        0      0      0             0      1000   -1           0
        0      0      0             0      0      5000         0
        0      0      0             0      0      1            -100;

Q =     0      -1000  0             0      0      0            0
        0      0      -1000         0      0      0            0
        0      0      0             -1200  0      0            0
        1000   0      0             0      0      0            100
        100    0      0             0      0      0            0
        0      0      0             0      0      0            2;

R =     0.1    10000  250           0      0      0
        0      0      0             0      20     10000
        0      0      0             1      0      0
        0      0      0             0      1      0;

F =     0      0      0      0      0      0.2    0;
ENDDATA

@FOR (ENVT(K): @FREE(G(K)));
@FOR(IMPACT(M): @FREE(H(M)));
@FOR(STREAM(I): @SUM(PROCESS(J): P(I,J)*S(J)) = F(I)); !Equation 7.1;
@FOR(ENVT(K): @SUM(PROCESS(J): Q(K,J)*S(J)) = G(K)); !Equation 7.3;
@FOR(IMPACT(M): @SUM(ENVT(K): R(M,K)*G(K)) = H(M)); !Equation 7.5;
```

A portion of the solution report obtained upon solving the model is shown below.

```
Feasible solution found.
  Infeasibilities:                                    0.000000
  Total solver iterations:                                   0
  Elapsed runtime seconds:                                0.12

  Model Class:                                              LP

  Total variables:                       17
  Nonlinear variables:                    0
  Integer variables:                      0

  Total constraints:                     17
  Nonlinear constraints:                  0

  Total nonzeros:                        40
  Nonlinear nonzeros:                     0

                                     Variable           Value
                                       S( P1)     0.2000300E-02
                                       S( P2)     0.2000000E-05
                                       S( P3)     0.1000000E-04
                                       S( P4)     0.1000150E-02
                                       S( P5)     0.2000000E-03
                                       S( P6)     0.2000000
                                       S( P7)     0.2000000E-02
                                       F( S1)      0.000000
                                       F( S2)      0.000000
                                       F( S3)      0.000000
                                       F( S4)      0.000000
                                       F( S5)      0.000000
                                       F( S6)      1000.000
                                       F( S7)      0.000000
                                       G( G1)    -0.2000000E-02
                                       G( G2)    -0.1000000E-01
                                       G( G3)     -1.200180
                                       G( G4)      2.200300
                                       G( G5)     0.2000300
                                       G( G6)     0.4000000E-02
                                       H( H1)     -400.0452
                                       H( H2)      44.00060
                                       H( H3)      2.200300
                                       H( H4)     0.2000300
```

The solution can easily be verified by examining Process 6 which corresponds to the use of the bulb. The desired functional unit is 1,000 h. However, Table 7.5 provides information on the generation of 5,000 h of incandescent bulb use. In order to achieve the desired 1,000 h functional unit, Process 6 will need to be scaled by 0.2 or the ratio 1,000/5,000 h. This corresponds to the scaling factor for Process 6 which is highlighted in the solution report above. Furthermore, the total resource depletion impact is equivalent to 400 RDU, eco-toxicity is equivalent to 44 m^2, global warming impact is 2.20 kg CO_2 eq., and the acidification impact is 0.20 kg SO_2 eq. The negative value of resource depletion indicates that the resources have been consumed by the system, which follows the convention adopted during system definition as indicated by the negative elements in Table 7.6. It is important to note that if the environmental impact evaluation was limited only in the examination of the use phase of the incandescent bulb (Process

6), it will appear that this activity does not have any environmental impact since all entries for Process 6 in Table 7.6 are equal to zero. The environmental impact is thus instead contributed by all other processes needed to perform the "lighting" function. This example clearly shows how a life cycle perspective reveals hidden or indirect environmental impacts.

For all parameters calculated in this case study, the solution is 100 times greater than the solution given by Heijungs [5] since Heijungs [5] made use of the functional unit equivalent to 10 h of incandescent bulb use instead of 1,000 h.

7.5 Conclusion

LCA is a rigorous methodology for the analysis of the environmental impacts of cradle-to-grave product systems. Such impacts account for both direct effects, as well as indirect impacts that occur in the background, away from the immediate point of use of a given technology. These impacts may also be geographically and temporally dispersed. Since a product life cycle system can be viewed as a "slice" or subset of a complete economic system, the general principles of calculation are similar to those of input–output analysis, but use slightly different conventions. In particular, the data convention is different because LCA typically uses bottom-up process data at a low level of aggregation (particularly for foreground processes), while conventional economic input–output models rely on a highly aggregated top-down approach. These two approaches are complementary and can even be combined into a hybrid approach as described by Heijungs and Suh [4].

References

1. Fava JA, Smerek A, Heinrich AB, Morrison L (2014) The role of the society of environmental toxicology and chemistry (SETAC) in life cycle assessment (LCA) development and application. In: Klopffer W (ed) Background and future prospects in life cycle assessment. Springer, Dordrecht, Netherlands, pp 39–83
2. Finkbeiner M (2014) The international standards as the constitution of life cycle assessment: the ISO 14040 series and its offspring. In: Klopffer W (ed) Background and future prospects in life cycle assessment. Springer, Dordrecht, Netherlands, pp 85–106
3. Heijungs R (1994) A generic method for the identification of options for cleaner products. A Generic Method Ecol Econ 10:69–81
4. Heijungs R, Suh S (2002) The computational structure of life cycle assessment. Springer, Dordrecht, Netherlands
5. Heijungs R (2009) An illustration of the LCA technique. In: Boersema JJ, Reijnders L (eds) Principles of environmental sciences. Springer, Dordrecht, Netherlands
6. Tan RR, Yu KDS, Aviso KB, Promentilla MAB (2017) Input-output modeling approach to sustainable systems engineering. In: Abraham M (ed) Encyclopedia of sustainable technology. Elsevier, Amsterdam, The Netherlands, pp 519–523

Chapter 8
Input–Output Models of Industrial Plants

Abstract Physical input–output models can be used to aid in the synthesis (design) or operations of industrial plant. In this chapter, the use of such models is illustrated for the case of polygeneration plants. The first example illustrates a zero degree of freedom synthesis problem. A second example illustrates how an input–output model can be used to identify process bottlenecks in existing plants. A third example discusses a mixed-integer linear programming (MILP) formulation of an input–output model to optimize state operations when the system is subjected to a disruptive event. LINGO code is provided for all examples.

Keywords Process synthesis · Debottlenecking · Abnormal operations
Mixed-integer linear programming

8.1 Introduction

The versatile I–O framework can also be used to develop models of at the level of the industrial plants. Such models are known as enterprise I–O models. Such models have been developed for a wide range of applications, including the production of steel [4], aluminum [5], coke [6], ceramic tiles [1], and dairy goods [3]. Other model variants have been developed for closed systems such as orbiting space stations [7] and multi-product systems such as polygeneration plants [9] and biorefineries [2]. The approach is generally applicable to a broad range of systems which satisfy the linearity assumptions used by these models.

8.2 Enterprise Input–Output Models

Enterprise I–O models differ from conventional economic input–output models in terms of both scale and calibration. Firstly, such models are limited in scale to a specific industrial plant site, although in some cases the approach is extended

R. R. Tan et al., *Input–Output Models for Sustainable Industrial Systems*,
Lecture Notes in Management and Industrial Engineering,
https://doi.org/10.1007/978-981-13-1873-3_8

loosely to industrial complexes or industrial parks. By comparison, conventional I–O models are usually applied to economic systems at the level of cities, countries, or regions. Secondly, the coefficients of enterprise I–O models are calibrated from engineering knowledge (i.e., first principles or records of stream flowrates) or from firm-level data; by comparison, conventional I–O models are based on economic statistics. Enterprise I–O models are often used to analyze interdependent process capacities within a system, an application which is analogous to the use of I–O models for the analysis of economic structures. Other demonstrated uses of these models include cost accounting [4], analysis of emissions to the environment [5], and process debottlenecking [9]. Further extensions of enterprise I–O models have also been developed for optimization purposes. Examples include optimal system design under uncertainty [7] and optimal adjustment to minimize losses under abnormal operating conditions [2].

The generic enterprise I–O model is:

$$\mathbf{Ax} = \mathbf{y} \tag{8.1}$$

where $\mathbf{A}$ is the $m \times n$ primary process matrix, $\mathbf{x}$ is the process throughput vector with n elements, and $\mathbf{y}$ is the primary plant output vector with m elements. Each row of the process matrix corresponds to a stream of a product or an intermediate, while each column corresponds to a process. In this case, the level of aggregation can vary depending on the application; thus, individual process units or clusters of processes can be represented as such in the model. Numerical values of the coefficients of $\mathbf{A}$ assume positive and negative values for output and input streams, respectively. Matrix $\mathbf{A}$ consists of column vectors $\mathbf{a}_j$, each representing a process within the system. The relative values of the elements in each column vector represent the ratios of stream flowrates, which are generally derived empirically or from first principles from an engineering perspective. Vector $\mathbf{y}$ is defined based on the expected demand of output from the process plant.

In the most basic case, it is necessary to solve for $\mathbf{x}$ given $\mathbf{A}$ and $\mathbf{y}$ using Eq. 8.1. The relative values of m and n, and Eq. 8.1 can give negative, zero, or positive degrees of freedom. Negative degrees of freedom $(m > n)$ indicate multi-functionality, as when some processes generate more than one output. In such cases, tight specifications on the values of some elements of $\mathbf{y}$ can be relaxed (i.e., replaced with inequalities or removed entirely) to allow a solution to be found. On the other hand, positive degrees of freedom $(m < n)$ indicate the presence of alternative processes, which allows for the potential to identify an optimal design of the system based on a suitable criterion. For systems with zero degrees of freedom $(m = n)$, given that $\mathbf{A}$ is nonsingular, the solution can be determined by matrix inversion:

$$\mathbf{x} = \mathbf{A}^{-1}\mathbf{y} \tag{8.2}$$

This step is similar to the process of determining sector output from final demand using the Leontief inverse in economic I–O models. Since $\mathbf{x}$ denotes the throughput of the processes within the plant, in the special case of static, steady-state operation, these flowrates also give the minimum rated capacities of the process units. The values in vector **x** are significant because these can be used to determine the flowrate of additional streams not included in **y**. Such streams can include inputs (e.g., resources, labor, feedstocks, raw materials, fuel) and outputs (e.g., wastes, pollutants, and byproducts whose flowrates are not previously specified). Following the convention described previously, positive and negative values denote net output and net input from the plant. The net flowrates of these streams can be determined as:

$$\mathbf{Bx} = \mathbf{z} \tag{8.3}$$

where **B** is the $p \times n$ secondary process matrix and **z** is the secondary plant output vector with p elements. The values in **B** follow the convention used for **A**. Every column vector $\mathbf{b}_j$ in **B** is a continuation of the corresponding vector $\mathbf{a}_j$ in **A**. Matrices **A** and **B** are thus linked by a common process throughput vector **x**. This linkage makes it possible to calculate the flowrates of the secondary streams (**z**) from the exogenously specified primary stream flowrates (**y**). Equations 8.2 and 8.3 can further be combined into simplified form:

$$\mathbf{BA}^{-1}\mathbf{y} = \mathbf{z} \tag{8.4}$$

In addition, this model formulation lends itself to the analysis of the profitability of a system. Revenues from the sale of products are a direct function of **y**, while the cost of purchasing various inputs can be determined from **z**. Meanwhile, the capital cost for each process unit can be estimated as a linear function of process capacities, given by **x**.

Thus, the profitability of the plant can be determined by:

$$P = \mathrm{AF}\left(\mathbf{c}_1^T\mathbf{y} + \mathbf{c}_2^T\mathbf{z}\right) - \mathrm{CRF}\mathbf{c}_3^T\mathbf{x} \tag{8.5}$$

where P is the annual profit, AF is a factor to convert all flowrates to annual basis, $\mathbf{c}_1$ is the unit price vector of the products in **y**, $\mathbf{c}_2$ is the unit price vector of the purchased inputs in **z** (with values of zero for elements corresponding to freely available natural resources and to pollutants discharged into the environment), CRF is the capital recovery factor (determined from discount rate and the economic life of the plant), and $\mathbf{c}_3$ is the variable cost coefficient of the linear cost functions of the process units. The last term of Eq. 8.5 assumes that the cost of the industrial plant is dominated by the capital costs of the major pieces of equipment and that all other capital costs are negligible. An alternative interpretation is that these other costs are embedded in the values of the elements of $\mathbf{c}_3$. Also, Eq. 8.5 can be augmented if needed by adding a fixed cost for each process unit.

8.3 Case Study 8.1

This example is adapted from the polygeneration case study in Sy et al. [8] and involves process synthesis or preliminary design. In such cases, the problem is to determine the capacities or throughputs of streams of component processes to give a balanced process system. A polygeneration system generates electricity along with other secondary products (e.g., heat, cooling, chemicals); such integrated systems are typically able to produce these outputs more efficiently than separate, stand-alone plants. The component units of such systems are interdependent due to their integration within the process. Their interactions can be modeled using the enterprise I–O approach. In this case, the system consists of four major processes, namely:

- A combined heat and power (CHP) module
- A boiler
- A vapor compression chiller
- A membrane-based water purification module.

The product streams of the polygeneration system are:

- Electricity (measured in kW)
- Heat as steam (measured in kW)
- Cooling as recirculating chilled water (measured in kW)
- Purified water (measured in L/h).

Other streams that are relevant for analysis purposes are:

- Fuel oil (measured in L/h)
- Feed water (measured in L/h)
- Rejected water (measured in L/h).

Note that the fuel oil and feed water streams are purchased inputs, while the rejected water stream is an effluent that is discharged by the plant. For this polygeneration system, the elements of **A** are given in Table 8.1, while the elements of **B** are given in Table 8.2. The individual process units are also illustrated in Fig. 8.1.

Given these performance data on the process units, we wish to specify the sizes or capacities of process units for a polygeneration plant whose output requirements are specified in the second column of Table 8.3. In other words, the problem is to find the vector $\mathbf{x}$, given the net output vector $\mathbf{y} = (10{,}000 \quad 25{,}000 \quad 8{,}000 \quad 100{,}000)^{\mathrm{T}}$. The fuel and freshwater requirement, as well as the generation of rejected wastewater, is also sought, as denoted by vector $\mathbf{z}$. In addition, the annual profit P of the system needs to be calculated from the revenues from sale of outputs minus the purchase of main inputs and the cost of capital recovery. The unit prices of the product and secondary streams ($\mathbf{c}_1$ and $\mathbf{c}_2$, respectively) are shown in the third column of Table 8.3. Note that the rejected water, which is a waste stream discharged to the environment, has no economic value. Capital cost coefficients are given in Table 8.4. It is assumed that the annual operating hours (AF) is 8000 h/year and the capital recovery factor (CRF) is 0.08.

Table 8.1 Elements of **A** for Case Study 8.1

	Process unit			
	CHP module	Boiler	Chiller	Membrane module
Electricity (kW)	1	−0.01	−0.2	−0.001
Heat (kW)	1.5	1	0	0
Cooling (kW)	0	0	1	0
Purified water (L/h)	−6.2	−2.1	0	1

Table 8.2 Elements of **B** for Case Study 8.1

	Process unit			
	CHP module	Boiler	Chiller	Membrane module
Fuel oil (L/h)	−0.32	−0.12	0	0
Freshwater (L/h)	0	0	0	−2.5
Rejected water (L/h)	0	0	0	1.5

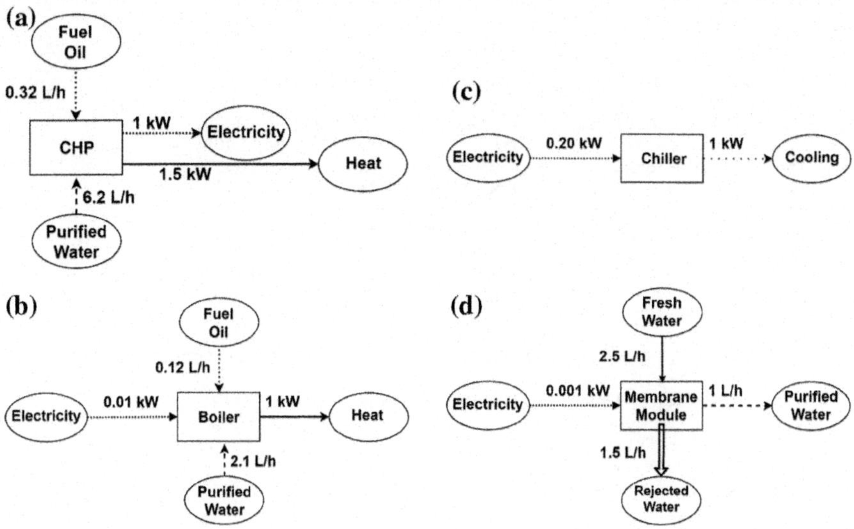

Fig. 8.1 Schematic diagram of process unit for **a** CHP, **b** boiler, **c** chiller, and **d** membrane module

Table 8.3 Elements of vectors **y**, **z**, $\mathbf{c}_1$, and $\mathbf{c}_2$ for Case Study 8.1

	Requirement	Unit Price
Electricity	10,000 kW	$0.16/kWh
Heat	25,000 kW	$0.05/kWh
Cooling	8,000 kW	$0.07/kWh
Purified water	100,000 L/h	$0.01/L
Fuel oil	?	$0.80/L
Freshwater	?	$0.001/L
Rejected water	?	$0/L

Table 8.4 Elements of $\mathbf{c}_3$ for Case Study 8.1

	Variable cost coefficient
CHP module	$950/kW
Boiler	$180/kW
Chiller	$270/kW
Membrane module	$14/(L/h)

The model can be determined by coding Eqs. 8.1, 8.3, and 8.5 in LINGO as follows:

```
!Equation 8.1;
1*x1        - 0.01*x2    - 0.2*x3     - 0.001*x4    = 10000;
1.5*x1      + 1*x2       - 0*x3       + 0*x4        = 25000;
0*x1        + 0*x2       + 1*x3       + 0*x4        = 8000;
-6.2*x1     - 2.1*x2     + 0*x3       + 1*x4        = 100000;

!Equation 8.3;
-0.32*x1    - 0.12*x2    + 0*x3       + 0*x4        = z1;
0*x1        + 0*x2       + 0*x3       - 2.5*x4      = z2;
0*x1        + 0*x2       + 0*x3       + 1.5*x4      = z3;

!Equation 8.5;
P = 8000*(0.16*10000 + 0.05*25000 + 0.07*8000 + 0.01*100000 + 0.8*z1 +
0.001*z2 + 0*z3)
- 0.08*(950*x1 + 180*x2 + 270*x3 + 14*x4);

!Additional constraints;
@free(z1);
@free(z2);
@free(P);
```

LINGO can then be prompted to find the solution to this system of equations to yield the following solution report:

Table 8.5 Process equipment requirements for Case Study 8.1

	Process unit			
	CHP module	Boiler	Chiller	Membrane module
Capacity	11,860.8 kW	7,208.9 kW	8,000 kW	188,675.3 L/h
Capital cost	$11,267,730	$1,2297,594	$2,160,000	$2,641,455

```
Feasible solution found.
  Infeasibilities:                                  0.000000
  Total solver iterations:                                 0

  Model Class:                                            LP

  Total variables:                       8
  Nonlinear variables:                   0
  Integer variables:                     0

  Total constraints:                     9
  Nonlinear constraints:                 0

  Total nonzeros:                       24
  Nonlinear nonzeros:                    0


                                        Variable           Value
                                              X1        11860.76
                                              X2        7208.854
                                              X3        8000.000
                                              X4        188675.3
                                              Z1       -4660.507
                                              Z2       -471688.3
                                              Z3        283013.0
                                               P        289907.1
```

The total annual profit (P) is found to be $289,907/year. The fuel oil and freshwater requirements are and 4,660 and 471,688 L/h, respectively, while 283,013 L/h of reject water is discharged from the plant. The required capacities and corresponding capital costs of the process units are given in Table 8.5. The cost estimate for the entire plant is considered as the sum of the capital costs of the four major process units, which is $17,366,770. Finally, the balanced flows for the system can be found by multiplying each column in Tables 8.1 and 8.2 by the corresponding element of vector **x**, results of which are summarized in Table 8.6 and illustrated in Fig. 8.2. Note that each row sum across the four process unit columns adds up to a net output corresponding to an element in vector **y** or **z**.

8.4 Case Study 8.2

The case study in the previous section is revisited here for a process debottlenecking problem. Such problems are characterized by an analysis of the performance of a previously existing process network under modified demand conditions.

Table 8.6 Balanced process table for Case Study 8.1

	CHP module	Boiler	Chiller	Membrane module	Net output
Electricity (kW)	11,861	−72	−1,600	−189	10,000
Heat (kW)	17,791	7,209	0	0	25,000
Cooling (kW)	0	0	8,000	0	8,000
Purified water (L/h)	−73,537	−15,139	0	188,675	100,000
Fuel oil (L/h)	−3,795	−865	0	0	−4,661
Freshwater (L/h)	0	0	0	−471,688	−471,688
Rejected water (L/h)	0	0	0	283,013	283,013

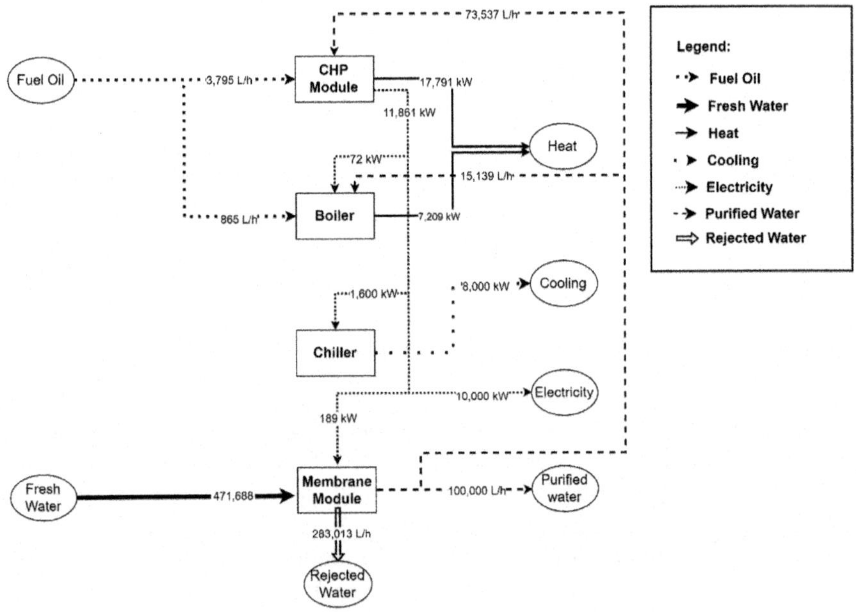

Fig. 8.2 Balanced flows of the system

First, it is assumed that the actual rated maximum capacities of the installed process units are much larger that the exact requirements calculated previously. There are two reasons for this. First, in most cases, commercially available process units are available in discrete sizes, instead of being custom-built for each specific plant. Thus, a larger unit can be installed and run below full load to meet the requirements. Secondly, it is common engineering practice to install excess capacity to allow for a margin of safety that can absorb future changes in demand, or deterioration of equipment performance with age. In this example, the installed capacities and the percentage safety margins are given in Table 8.7.

Table 8.7 Process equipment requirements for Case Study 8.1

	Process unit			
	CHP module	Boiler	Chiller	Membrane module
Required capacity	11,860.8 kW	7,208.9 kW	8,000 kW	188,675.3 L/h
Installed capacity	12,000 kW	8,000 kW	10,000 kW	200,000 L/h
Capacity margin	1.2%	11%	25%	6%

Let us assume that the polygeneration plant is built with these process unit capacities. Furthermore, suppose that at some point during the operating life of the plant, the demand for cooling increases from the original 8,000 to 9,000 kW. This is an increase of 12.5% and is well within the capacity limit of the installed chiller. However, bottlenecks can occur elsewhere in the plant due to the interdependencies among the process units. Any resultant bottlenecks can be found by repeating the same procedure used in Case Study 8.1, but with additional constraints restricting the elements of vector **x** to values up to the installed capacities in Table 8.7. The revised LINGO model is:

```
!Equation 8.1;
1*x1        - 0.01*x2    - 0.2*x3     - 0.001*x4    = 10000;
1.5*x1      + 1*x2       - 0*x3       + 0*x4        = 25000;
0*x1        + 0*x2       + 1*x3       + 0*x4        = 9000;
-6.2*x1     - 2.1*x2     + 0*x3       + 1*x4        = 100000;

!Equation 8.3;
-0.32*x1    - 0.12*x2    + 0*x3       + 0*x4        = z1;
0*x1        + 0*x2       + 0*x3       - 2.5*x4      = z2;
0*x1        + 0*x2       + 0*x3       + 1.5*x4      = z3;

!Additional constraints;
@free(z1);
@free(z2);

!Installed capacity limits;
x1 < 12000;
x2 < 8000;
x3 < 10000;
x4 < 200000;
```

Attempting to solve this model using LINGO leads to an infeasible solution as shown in the following solution report:

```
No feasible solution found.
  Infeasibilities:                                     58.40209
  Total solver iterations:                                    4

  Model Class:                                               LP

  Total variables:                      7
  Nonlinear variables:                  0
  Integer variables:                    0

  Total constraints:                   12
  Nonlinear constraints:                0

  Total nonzeros:                      21
  Nonlinear nonzeros:                   0

                                          Variable          Value
                                                X1       12058.40
                                                X2       6912.397
                                                X3       9000.000
                                                X4       189278.1
                                                Z1      -4688.176
                                                Z2      -473195.3
                                                Z3       283917.2
```

Inspection of this solution report reveals that the new required capacity of the CHP module is 12,058.4 kW, which exceeds the 12,000 kW rated capacity of the installed unit. This result indicates that the plant will be unable to meet the additional 12.5% cooling load, because of the extra electricity demand that the chiller will draw from the CHP module. Thus, the latter process unit is identified as the process bottleneck. Engineering interventions can then be made to eliminate the bottleneck, such as retrofit, process intensification, replacement, or installation of a supplementary process module. The detailed discussion of such options is beyond the scope of this chapter.

8.5 Case Study 8.3

In addition to process synthesis and debottlenecking, the enterprise I–O model can is also extended to applications involving optimization of abnormal operations in industrial plants. In such cases, disruptions due to equipment failure or maintenance, or due to shortage of critical inputs, can cause a plant to be partially inoperable. In such cases, it will be of significant interest to control economic losses by determining the best temporary operating state for the duration of the crisis [2]. Such a strategy will require sacrificing some product streams and prioritizing others. The mixed-integer linear programming (MILP) developed by Kasivisvanathan et al. [2] is presented here to deal with such problems. The objective is to maximize the net revenues:

$$\max \quad \mathbf{c}_1^T \mathbf{y} * + \mathbf{c}_2^T \mathbf{z}^* \tag{8.6}$$

where $\mathbf{y}^*$ and $\mathbf{z}^*$ are the flowrates of the products and secondary streams during abnormal operations. The objective function is similar to Eq. 8.5, but the term for capital cost no longer appears since it is assumed that the plant has already been built and the corresponding costs have already been incurred. The stream balances are given by:

$$\mathbf{A}\mathbf{x}^* = \mathbf{y}^* \tag{8.7}$$

$$\mathbf{B}\mathbf{x}^* = \mathbf{z}^* \tag{8.8}$$

where $\mathbf{x}^*$ signifies the throughput or utilized capacity of the process units during abnormal operations. It can be seen that these constraints are similar in form to Eqs. 8.1 and 8.3. The vectors signifying flowrates are subject to constraints due to the abnormal conditions:

$$\mathbf{s}^T\mathbf{x}_{min} \leq \mathbf{x}^* \leq \mathbf{s}^T\mathbf{x}_{max} \tag{8.9}$$

where $\mathbf{x}_{max}$ is the maximum limiting throughput of the process units, $\mathbf{x}_{min}$ is the minimum part load operating throughput below which a process unit becomes unstable, and $\mathbf{s}$ is the vector of binary variables indicating whether a process unit is switched on or off during abnormal operations. The upper limits may be the rated capacities of the units themselves (as previously discussed, there will often be excess capacity); alternatively, the values may be exogenously defined if a process unit becomes partially inoperable due to damage. In such cases, the limit is defined by the highest capacity possible in this abnormal state. The flowrates of product streams are limited as follows:

$$\mathbf{y}_{min} \leq \mathbf{y}^* \leq \mathbf{y}_{max} \tag{8.10}$$

where $\mathbf{y}_{min}$ and $\mathbf{y}_{max}$ are the lower and upper limits for the outputs, as imposed by customer requirements, during the crisis. Equation 8.11 ensures that the flexibility of allocating capacity during abnormal operations is not completely under the control of plant management, but is also subject to external conditions. Likewise, the flowrates of the secondary streams are subject to similar limits:

$$\mathbf{z}_{min} \leq \mathbf{z}^* \leq \mathbf{z}_{max} \tag{8.11}$$

where $\mathbf{z}_{min}$ and $\mathbf{z}_{max}$ are the lower and upper limits of the secondary stream flowrates during the abnormal operations. This constraint can also be used to limit flowrates of critical process inputs if the crisis is due to such a shortage.

The polygeneration plant in Case Study 8.2 is used to illustrate the use of this model. It is assumed that the crisis is caused by a water shortage, which limits freshwater availability to 300,000 L/h. The operating ranges of the process equipment are summarized in Table 8.8. The process units can be adjusted to operate only within these ranges and can be switched off otherwise. The required ranges of net outputs of the four product streams are given in Table 8.9.

Table 8.8 Process equipment requirements for Case Study 8.1

	Process unit			
	CHP module	Boiler	Chiller	Membrane module
Installed capacity	12,000 kW	8,000 kW	10,000 kW	200,000 L/h
Minimum capacity	4,000 kW	2,000 kW	4,000 kW	0 L/h

Table 8.9 Elements of vectors **y**, **z**, $\mathbf{c}_1$, and $\mathbf{c}_2$ for Case Study 8.1

	Minimum requirement	Maximum requirement
Electricity	5,000 kW	10,000 kW
Heat	5,000 kW	25,000 kW
Cooling	2,000 kW	8,000 kW
Purified water	50,000 L/h	100,000 L/h

The corresponding LINGO model can be coded as shown below. Note that in Eq. 8.11, arbitrarily large negative or positive numbers can be used to signify if there are no practical limits to the flowrates of the secondary streams. Alternatively, these constraints can be removed from the model.

```
!Equation 8.6;

max = (0.16*10000 + 0.05*25000 + 0.07*8000 + 0.01*100000 + 0.8*z1 + 0.001*z2
+ 0*z3);

!Equation 8.7;
1*x1        - 0.01*x2     - 0.2*x3     - 0.001*x4    = y1;
1.5*x1      + 1*x2        - 0*x3       + 0*x4        = y2;
0*x1        + 0*x2        + 1*x3       + 0*x4        = y3;
-6.2*x1     - 2.1*x2      + 0*x3       + 1*x4        = y4;

!Equation 8.8;
-0.32*x1    - 0.12*x2     + 0*x3       + 0*x4        = z1;
0*x1        + 0*x2        + 0*x3       - 2.5*x4      = z2;
0*x1        + 0*x2        + 0*x3       + 1.5*x4      = z3;

!Equation 8.9;
4000*s1 < x1;       x1 <  12000*s1;    @bin(s1);
2000*s2 < x2;       x2 <  8000*s2;     @bin(s2);
4000*s3 < x3;       x3 <  10000*s3;    @bin(s3);
0*s4 < x4;          x4 <  200000*s4;   @bin(s4);

!Equation 8.10;
5000 < y1;          y1 < 10000;
5000 < y2;          y2 < 25000;
2000 < y3;          y3 < 10000;
50000 < y4;         y4 < 200000;

!Equation 8.11;
-10^6 < z1;         z1 < 0;
-300000 < z2;       z2 < 0;
0 < z3;             z3 < 10^6;

!Additional constraints;
@free(z1);
@free(z2);
```

Solving this MILP gives the LINGO solution report below. This linear model can be readily solved to global optimality with no computational difficulties.

```
Global optimal solution found.
  Objective value:                              2686.816
  Objective bound:                              2686.816
  Infeasibilities:                              0.000000
  Extended solver steps:                               0
  Total solver iterations:                             2

  Model Class:                                      MILP

  Total variables:                 15
  Nonlinear variables:              0
  Integer variables:                4

  Total constraints:               25
  Nonlinear constraints:            0

  Total nonzeros:                  47
  Nonlinear nonzeros:               0

                              Variable           Value        Reduced Cost
                                    Z1       -1883.679            0.000000
                                    Z2       -216240.7            0.000000
                                    Z3        129744.4            0.000000
                                    X1        5886.496            0.000000
                                    X2        0.000000           0.1045556
                                    X3        4000.000            0.000000
                                    X4        86496.28            0.000000
                                    Y1        5000.000            0.000000
                                    Y2        8829.744            0.000000
                                    Y3        4000.000            0.000000
                                    Y4        50000.00            0.000000
                                    S1        1.000000            0.000000
                                    S2        0.000000            0.000000
                                    S3        1.000000            218.5550
                                    S4        1.000000            0.000000
```

The optimal operating states of the process units under the specified abnormal conditions are given in Table 8.10. Note that this solution recommends switching off the boiler entirely and running the other three process units at part load. The objective function value is $2,686.82/year. This figure corresponds to the difference between revenues from sales of the four products minus the purchase cost of the fuel and freshwater inputs. If the capital recovery cost is factored in, the process actually runs at a loss during this abnormal state. However, this solution represents the best "damage control" strategy to minimize economic losses. The balanced

Table 8.10 Optimal operating state of the process units for Case Study 8.3

	Operating state
CHP module	5,886 kW
Boiler	0 kW
Chiller	4,000 kW
Membrane module	86,496 L/h

Table 8.11 Optimal balanced process table for Case Study 8.3

	CHP module	Boiler	Chiller	Membrane module	Net output
Electricity (kW)	5,886	0	−800	−86	5,000
Heat (kW)	8,830	0	0	0	8,830
Cooling (kW)	0	0	4,000	0	4,000
Purified water (L/h)	−36,496	0	0	86,496	50,000
Fuel oil (L/h)	−1,884	0	0	0	−1,884
Freshwater (L/h)	0	0	0	−216,241	−216,241
Rejected water (L/h)	0	0	0	129,744	129,744

process table for the abnormal state can be determined using the same procedure as in Case Study 8.1 and is shown in Table 8.11 and Fig. 8.3. Note that the water requirement is 216,241 L/h, which is well within the 300,000 L/h limit specified above.

If needed, LINGO can also generate near-optimal solutions with alternative topologies. This feature can be activated by clicking the "LINGO" in the toolbar of the main interface, selecting "Options," and then clicking the "Integer Solver" tab. At the bottom of the dialog box, under "*k*-best," the desired number of alternative solutions can be selected. If $k = 2$, the second best solution can be displayed as in the report below.

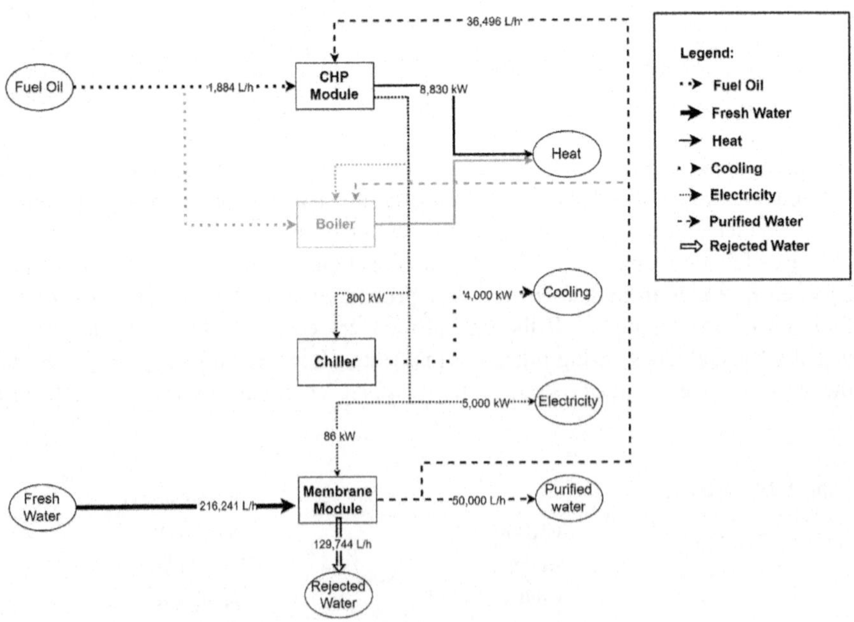

Fig. 8.3 Optimal operational adjustment for Case Study 8.3

```
Global optimal solution found.
  Objective value:                              2686.816
  Objective bound:                              2686.816
  Infeasibilities:                              0.000000
  Extended solver steps:                               0
  Total solver iterations:                             2

  Global optimal solution found.
  Objective value:                              2477.705
  Objective bound:                              2477.705
  Infeasibilities:                              0.000000
  Extended solver steps:                               0
  Total solver iterations:                             0

  Model Class:                                      MILP

  Total variables:                   0
  Nonlinear variables:               0
  Integer variables:                 0

  Total constraints:                 0
  Nonlinear constraints:             0

  Total nonzeros:                    0
  Nonlinear nonzeros:                0


                                Variable           Value
                                      Z1       -2131.471
                                      Z2       -227118.1
                                      Z3        136270.9
                                      X1        5910.847
                                      X2        2000.000
                                      X3        4000.000
                                      X4        90847.25
                                      Y1        5000.000
                                      Y2        10866.27
                                      Y3        4000.000
                                      Y4        50000.00
                                      S1        1.000000
                                      S2        1.000000
                                      S3        1.000000
                                      S4        1.000000
```

The objective function value for the second best solution is \$2477.71/year, which gives an additional \$209.11/year in losses, as compared to the previously reported optimal solution. In this solution, all four process units remain switched on during abnormal operating conditions, as shown in Table 8.12. This option results in a different allocation of streams, as well as net plant output, as shown in the balanced process matrix in Table 8.13 with the schematic diagram equivalent in Fig. 8.4.

Table 8.12 Alternative near-optimal operating state of the process units for Case Study 8.3

	Operating state
CHP module	5,911 kW
Boiler	2,000 kW
Chiller	4,000 kW
Membrane module	90,847 L/h

Table 8.13 Alternative near-optimal balanced process table for Case Study 8.3

	CHP module	Boiler	Chiller	Membrane module	Net output
Electricity (kW)	5,911	−20	−800	−91	5,000
Heat (kW)	8,866	2,000	0	0	10,866
Cooling (kW)	0	0	4,000	0	4,000
Purified water (L/h)	−36,647	−4,200	0	90,847	50,000
Fuel oil (L/h)	−1,891	−240	0	0	−2,131
Freshwater (L/h)	0	0	0	−227,118	−227,118
Rejected water (L/h)	0	0	0	136,271	136,271

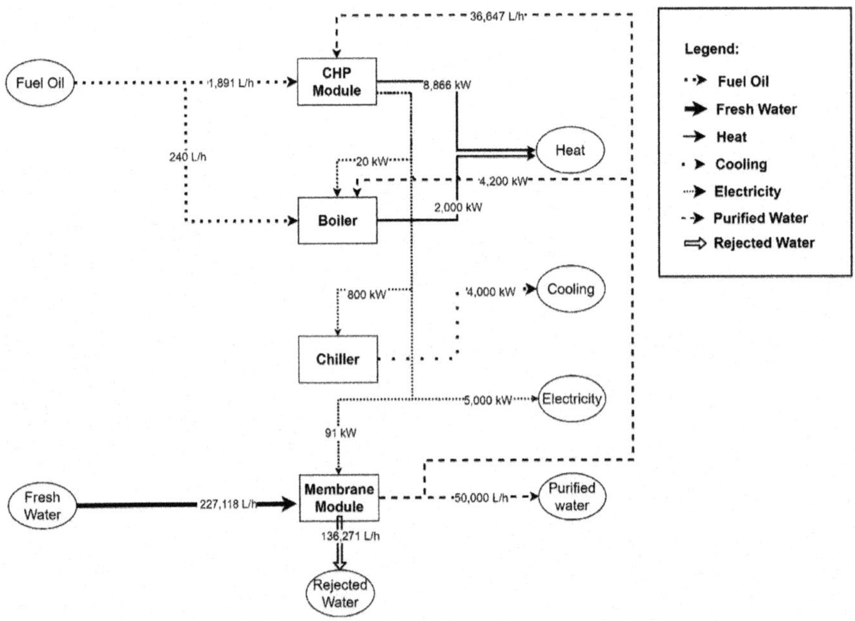

Fig. 8.4 Near-optimal solution for Case Study 8.3

8.6 Conclusion

Enterprise I–O models can be used at the level of industrial plants. Such models can be used for preliminary design or synthesis, which involves determining the capacities of individual component processes to achieve a balanced system. Variants of the model can also be applied for existing plants. For example, process debottlenecking can be done with such models to facilitate planning for capacity expansion in response to changes in product demand. Other model extensions can be used to determine optimal adjustments for operations under abnormal conditions, which may be brought about by equipment failure or by a shortage of inputs.

References

1. Albino V, Kuhtz S (2004) Enterprise input–output model for local sustainable development—the case of a tiles manufacturer in Italy. Resour Conserv Recycl 41:165–176
2. Kasivisvanathan H, Barilea IDU, Ng DKS, Tan RR (2013) Optimal operational adjustment in multi-functional energy systems in response to process inoperability. Appl Energy 102:492–500
3. Lenzen M, Lundie S (2012) Constructing enterprise input-output tables—a case study of New Zealand dairy products. J Econ Struct, 1, Article 6
4. Lin X, Polenske KR (1998) Input–output modeling of production processes for business management. Struct Change Econ Dyn 9:205–226
5. Lupis CHP, McMichael FC (1999) Input–output microeconomic model for process and environmental analysis. In Proceedings of the TMS fall extraction and processing conference, vol 3, pp 1905–1918
6. Polenske KR, McMichael FC (2002) A Chinese cokemaking process-flow model for energy and environmental analyses. Energy Policy 30:865–883
7. Quirk J, Olson M, Habib-Agahi H, Fox G (1989) Uncertainty and Leontief systems: an application to the selection of space station system designs. Manage Sci 35:585–596
8. Sy CL, Aviso KB, Ubando AT, Tan RR (2018) Synthesis of cogeneration, trigeneration, and polygeneration systems using target-oriented robust optimization. In: De S, Bandyopadhyay S, Assadi M, Mukherjee DA (eds) Sustainable energy technology and policies., Springer, pp 155–171
9. Tan RR, Lam HL, Kasivisvanathan H, Ng DKS, Foo DCY, Kamal M, Hallale N, Klemeš JJ (2012) An algebraic approach to identifying bottlenecks in linear process models of multifunctional energy systems. Theor Found Chem Eng 46:642–650

Chapter 9
Input–Output Models of Organizational Structures

Abstract The interactions among groups of employees or staff in organizational units involve transactions that are analogous to those reflected in large-scale input–output models. The main difference is that, in an organization, the interactions of personnel with internal or external customers are often measured in terms of time (e.g., person-hours), in contrast to the monetary and physical units used in traditional input–output models. This chapter discusses organizational input–output models based on this concept. A case study is solved under different scenarios to illustrate the framework, and the corresponding LINGO model formulations are given.

Keywords Human resources · Organizational input–output model
Workforce allocation

9.1 Introduction

Human resource planning and management are essential in the operation of any organization. It includes evaluating competencies and forecasting workforce needs when changes in the external demand are expected. The information is then used for estimating needs to meet business targets or for planning human resource recruitment in the interest of business expansion and in preparation for employee retirement. In addition, it has been recognized that human resources are vital in the continuous operation of organizations especially during adverse events [7]. Some strategies make use of expert judgment, while others may implement a more analytic approach in finding the solution. This chapter shows how the input–output framework can be used for human resource planning in organizations and how it can be implemented in the optimization software LINGO.

R. R. Tan et al., *Input–Output Models for Sustainable Industrial Systems*,
Lecture Notes in Management and Industrial Engineering,
https://doi.org/10.1007/978-981-13-1873-3_9

9.2 Organizational Structures

Organizational structures can be represented as a system with interdependent components where the outputs of some components become inputs of another. In an open system, the components may be classified as either internal or external. External components are those which provide primary inputs into the system which do not directly translate into the external output of the organization. These can also be components which receive inputs from internal components but do not provide output used by the internal components. Examples of external components include clients or the public who are indirect sources of resource to an organization and the government which regulates and affects organizational operations. Internal components on the other hand refer to the interdependent units of an organization and may be represented by departmental units or job classifications. The organizational structure of a firm provides information on how these classifications can be made. It is important to note however that a clear distinction between classifications is done such that there are no overlaps and the categories are exhaustive [4]. The interdependencies between system components represent the nature by which they interact with each other. These interactions are essentially small-scale economic flows of services to internal customers, but are measured in person-hours rather than in terms of actual economic value. An input–output table is constructed based on the components identified and the intensity of interaction between them. The intensity may be quantified according to the number of persons in each classification and the number of hours spent by each person when interacting with others.

Correa and Craft [4] proposed the equivalent organizational input–output model for an organization with n components as indicated in Eq. 9.1.

$$\mathbf{At} + \mathbf{e} = \mathbf{t} \tag{9.1}$$

where $\mathbf{A}$ is an $n \times n$ workforce interaction matrix, containing elements a_{ij} as defined by Eq. 9.2, $\mathbf{e}$ is the external demand vector or net output of work, and $\mathbf{t}$ is total workload vector.

$$a_{ij} = \frac{P_{ij}}{t_j} \tag{9.2}$$

P_{ij} refers to the elements in transaction matrix $\mathbf{P}$ that corresponds to the number of personnel hours allocated by component i to provide services for component j, while t_j refers to the total number of personnel hours available in component j and is equivalent to the number of persons assigned in category j. This assumes that transaction matrix $\mathbf{P}$ represents the ideal state of organizational operation and that matrix $\mathbf{A}$ reflects the intensity of interaction between organizational components. Any change in the external demand, such as an increase in the need for an organization's services, is reflected by an increase in the elements of matrix $\mathbf{e}$. Thus, a change in human resource requirement can be calculated by looking for vector **t** using Eqs. 9.3 and 9.4 where **I** is an identity matrix.

$$\mathbf{e} = (\mathbf{I} - \mathbf{A})\mathbf{t} \tag{9.3}$$

$$\mathbf{t} = (\mathbf{I} - \mathbf{A})^{-1}\mathbf{e} \tag{9.4}$$

This framework has been used for analyzing workforce requirements in hospitals [6], libraries [3], and government offices [5]. The framework was also extended to integrate degrees of freedom for human resource allocation during crisis conditions using fuzzy optimization [1] and graph-theoretical approaches [2].

Unlike the traditional applications of the organizational I–O, which focused more on projecting workforce needs when changes in external demand occur, crisis conditions look at how organizations can maximize their productivity given the reduced workforce. Workforce availability is often affected due to disruptions caused by extreme events because of decreased mobility and increased incidence of sickness to name a few. The next section demonstrates how the organizational I–O model can be implemented beginning with a motivating example and a case study on how the model can be used for (1) determining workforce needs during changes in external demand and (2) optimal workforce allocation during crisis conditions.

9.3 Motivating Example

To demonstrate how the organizational I–O model can be implemented, a simple motivating example is discussed which was adapted from the work of Correa and Craft [4]. Take, for example, a small organization consisting of 750 employees which are distributed in four different departments—(1) high management, (2) middle management, (3) administration (admin), and (4) production (product). These departments interact with each other in order to perform their respective tasks. Based on observation, personnel utilization is as given in Table 9.1 with the equivalent schematic diagram presented in Fig. 9.1. The last column of Table 9.1 indicates the total number of personnel working in the department specified in the row label. For this hypothetical firm, only high management and administration interact with the external clients. Table 9.1 corresponds to the transaction matrix, **P**, described in Eq. 9.2.

The total workload (**t**) and external demand (**e**) vectors can be extracted from Table 9.1 as follows:

$$\mathbf{t} = \begin{bmatrix} 50 \\ 55 \\ 313 \\ 332 \end{bmatrix} \quad \mathbf{e} = \begin{bmatrix} 22 \\ 0 \\ 118 \\ 0 \end{bmatrix}$$

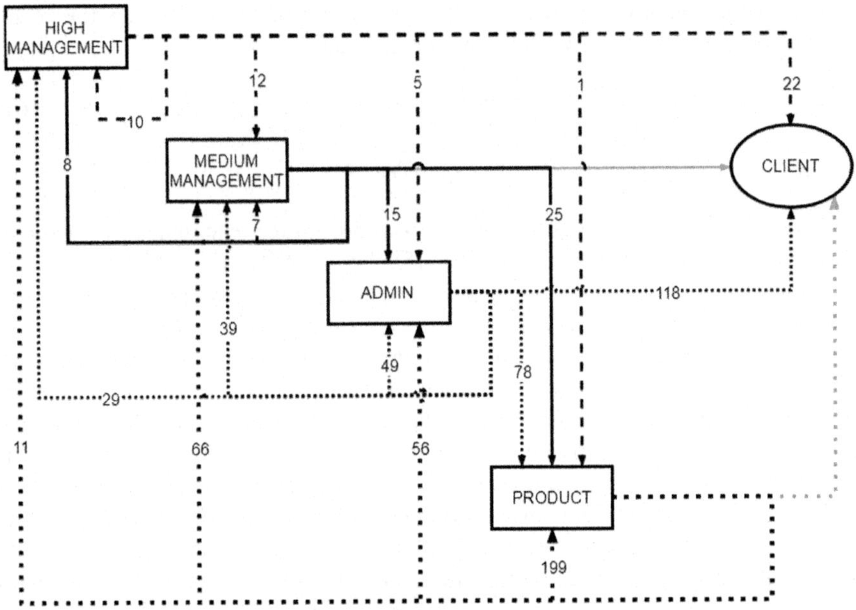

Fig. 9.1 Utilization of personnel in hypothetical firm [4]

Table 9.1 Utilization of firm's personnel in number of persons

	High management	Medium management	Administration	Production	External clients	Total
High management	10	12	5	1	22	50
Medium management	8	7	15	25	0	55
Administration	29	39	49	78	118	313
Production	11	66	56	199	0	332

Adapted from [4]

Similarly, the transactions in Table 9.1 can easily be translated to matrix **A** to reflect the intensity of interaction between departments using Eq. 9.2.

$$\mathbf{A} = \begin{bmatrix} 10/50 & 12/55 & 5/313 & 1/332 \\ 8/50 & 7/55 & 15/313 & 25/332 \\ 29/50 & 39/55 & 49/313 & 78/332 \\ 11/50 & 66/55 & 56/313 & 199/332 \end{bmatrix} = \begin{bmatrix} 0.20 & 0.22 & 0.02 & 0.003 \\ 0.16 & 0.13 & 0.05 & 0.075 \\ 0.58 & 0.71 & 0.16 & 0.235 \\ 0.22 & 1.20 & 0.18 & 0.599 \end{bmatrix}$$

Using Eq. 9.1 then results in the expression below:

$$\begin{bmatrix} 0.20 & 0.22 & 0.02 & 0.003 \\ 0.16 & 0.13 & 0.05 & 0.075 \\ 0.58 & 0.71 & 0.16 & 0.235 \\ 0.22 & 1.20 & 0.18 & 0.599 \end{bmatrix} \begin{bmatrix} 50 \\ 55 \\ 313 \\ 332 \end{bmatrix} + \begin{bmatrix} 22 \\ 0 \\ 118 \\ 0 \end{bmatrix} = \begin{bmatrix} 50 \\ 55 \\ 313 \\ 332 \end{bmatrix}$$

$$\underbrace{\begin{bmatrix} 28 \\ 55 \\ 195 \\ 332 \end{bmatrix}}_{\text{Intermediate demand}} + \underbrace{\begin{bmatrix} 22 \\ 0 \\ 118 \\ 0 \end{bmatrix}}_{\text{External demand}} = \underbrace{\begin{bmatrix} 50 \\ 55 \\ 313 \\ 332 \end{bmatrix}}_{\text{Total workload}}$$

$(\mathbf{I} - \mathbf{A})$ can then be calculated and its corresponding inverse $(\mathbf{I} - \mathbf{A})^{-1}$ as shown below.

$$(\mathbf{I} - \mathbf{A}) = \begin{bmatrix} 1 & 0 & 0 & 0 \\ 0 & 1 & 0 & 0 \\ 0 & 0 & 1 & 0 \\ 0 & 0 & 0 & 1 \end{bmatrix} - \begin{bmatrix} 0.20 & 0.22 & 0.02 & 0.003 \\ 0.16 & 0.13 & 0.05 & 0.075 \\ 0.58 & 0.71 & 0.16 & 0.235 \\ 0.22 & 1.20 & 0.18 & 0.599 \end{bmatrix}$$

$$= \begin{bmatrix} 0.80 & -0.22 & -0.02 & -0.003 \\ -0.16 & 0.87 & -0.05 & -0.075 \\ -0.58 & -0.71 & 0.84 & -0.235 \\ -0.22 & -1.20 & -0.18 & 0.400 \end{bmatrix}$$

$$(\mathbf{I} - \mathbf{A})^{-1} = \begin{bmatrix} 1.58 & 0.84 & 0.13 & 0.25 \\ 0.90 & 2.51 & 0.30 & 0.65 \\ 3.23 & 5.61 & 2.05 & 2.28 \\ 5.00 & 10.49 & 1.88 & 5.61 \end{bmatrix}$$

Furthermore, to determine any change in personnel requirement ($\mathbf{t}$) resulting from a change in external demand ($\mathbf{e}$), the expression below which resembles Eq. 9.4 can then be used.

$$\mathbf{t} = \begin{bmatrix} 1.58 & 0.84 & 0.13 & 0.25 \\ 0.90 & 2.51 & 0.30 & 0.65 \\ 3.23 & 5.61 & 2.05 & 2.28 \\ 5.00 & 10.49 & 1.88 & 5.61 \end{bmatrix} \mathbf{e}$$

9.4 Case Study 9.1

This case study is based on the example presented in Correa and Parker [6] which involves an acute care hospital with 563 full-time employees divided into ten departments. The description of the ten departments is shown in Table 9.2, while the transaction matrix (**P**) is shown in Table 9.3. The last column of the transaction matrix shows the number of employees assigned in a specific department such that the sum of the last column is equal to 563. The largest department is that which houses the nursing staff. There are currently 198 employees in this department. The smallest department on the other hand belongs to the category of high-level management which currently has four employees. Furthermore, the entries in Table 9.3 indicate the number of man-days per day spent by employees to provide services for external ($\mathbf{e}_0$) and internal clients.

Table 9.3 can be translated into the workforce interaction matrix **A** using Eq. 9.2. Take, for example, the entry in the third row, second column, this represents the amount of time spent by middle management (Department 2) to provide services for the high-level management (Department 1). This corresponds to the entry $P_{2,1} = 3.60$ which indicates that 3.60 man-days per day are required. The equivalent entry in the workforce interaction matrix **A** is to calculate the equivalent man-days per day needed per unit output of middle management ($a_{2,1} = 3.60/4 = 0.90$). All other entries in the matrix can be obtained in a similar way, and it is assumed that the organization will function according to these fixed ratios of interaction. The completed **A** matrix is shown in Table 9.4.

Using matrix **A**, the resulting values of matrix ($\mathbf{I} - \mathbf{A}$) are given in Table 9.5.

Table 9.2 List of departments in acute care hospital

	Departments
1	High-level management
2	Middle-level management
3	General administration
4	Support administration
5	Finance administration
6	Human services
7	Information services
8	Medical staff
9	Nursing staff
10	Ancillary staff

Table 9.3 Transaction matrix of hypothetical acute care hospital in man-days per day [6]

Dept.	1	2	3	4	5	6	7	8	9	10	$\mathbf{e_0}$	**t**
1	0.80	1.40	0.04	0.04	0.04	0.04	0.20	0.04	0.40	0.20	0.80	4
2	3.60	2.40	1.50	0.30	1.50	2.40	1.50	4.80	4.50	4.50	3.00	30
3	0.83	20.75	24.90	0.83	4.15	4.15	1.66	5.81	1.66	12.45	5.81	83
4	0.12	0.12	0.00	1.20	0.12	0.36	0.60	1.20	2.40	3.60	2.28	12
5	4.32	6.24	0.96	0.96	19.20	0.96	3.84	2.40	0.96	0.96	7.20	48
6	0.60	1.20	0.60	0.60	0.60	2.40	0.60	0.60	0.60	1.80	2.40	12
7	5.55	11.10	5.55	1.11	16.65	5.55	33.30	16.65	5.55	7.77	2.22	111
8	0.45	0.45	0.45	0.45	0.45	0.45	0.45	1.35	4.50	13.50	22.50	45
9	1.98	1.98	1.98	1.98	1.98	1.98	1.98	3.96	29.70	19.80	130.68	198
10	0.20	1.00	1.40	0.20	0.80	1.00	1.40	6.60	1.00	2.40	4.00	20

Table 9.4 Workforce interaction matrix **A**

Dept.	1	2	3	4	5	6	7	8	9	10
1	0.200	0.047	0.000	0.003	0.001	0.003	0.002	0.001	0.002	0.010
2	0.900	0.080	0.018	0.025	0.031	0.200	0.014	0.107	0.023	0.225
3	0.208	0.692	0.300	0.069	0.086	0.346	0.015	0.129	0.008	0.623
4	0.030	0.004	0.000	0.100	0.003	0.030	0.005	0.027	0.012	0.180
5	1.080	0.208	0.012	0.080	0.400	0.080	0.035	0.053	0.005	0.048
6	0.150	0.040	0.007	0.050	0.013	0.200	0.005	0.013	0.003	0.090
7	1.388	0.370	0.067	0.093	0.347	0.463	0.300	0.370	0.028	0.389
8	0.113	0.015	0.005	0.038	0.009	0.038	0.004	0.030	0.023	0.675
9	0.495	0.066	0.024	0.165	0.041	0.165	0.018	0.088	0.150	0.990
10	0.050	0.033	0.017	0.017	0.017	0.083	0.013	0.147	0.005	0.120

Table 9.5 (**I** − **A**) matrix

Dept.	1	2	3	4	5	6	7	8	9	10
1	0.800	−0.047	0.000	−0.003	−0.001	−0.003	−0.002	−0.001	−0.002	−0.010
2	−0.900	0.920	−0.018	−0.025	−0.031	−0.200	−0.014	−0.107	−0.023	−0.225
3	−0.208	−0.692	0.700	−0.069	−0.086	−0.346	−0.015	−0.129	−0.008	−0.623
4	−0.030	−0.004	0.000	0.900	−0.003	−0.030	−0.005	−0.027	−0.012	−0.180
5	−1.080	−0.208	−0.012	−0.080	0.600	−0.080	−0.035	−0.053	−0.005	−0.048
6	−0.150	−0.040	−0.007	−0.050	−0.013	0.800	−0.005	−0.013	−0.003	−0.090
7	−1.388	−0.370	−0.067	−0.093	−0.347	−0.463	0.700	−0.370	−0.028	−0.389
8	−0.113	−0.015	−0.005	−0.038	−0.009	−0.038	−0.004	0.970	−0.023	−0.675
9	−0.495	−0.066	−0.024	−0.165	−0.041	−0.165	−0.018	−0.088	0.850	−0.990
10	−0.050	−0.033	−0.017	−0.017	−0.017	−0.083	−0.013	−0.147	−0.005	0.880

Given these initial information, we will consider two different scenarios of the case study. Scenario 1 deals with forecasting human resource needs when the external demand changes, while Scenario 2 considers the optimal allocation of human resources during a crisis condition.

9.4.1 Scenario 1

The acute care hospital expects a change in the external demand for medical and nursing staff such that the demand from the medical staff is reduced by 5 units such that instead of 22.50 the demand becomes 17.50. The demand for nursing staff on another hand increased by 5 units from 130.68 to 135.68. The new external demand vector then becomes $\mathbf{e} = (0.80\ 3.00\ 5.81\ 2.28\ 7.20\ 2.40\ 2.22\ 17.50\ 135.68\ 4.00)^T$. This change demonstrates an increase in demand for the services of the nursing staff department and a proportional decrease in the external demand from the medical staff department. The objective is to determine the total amount of workforce required to meet this change in external demand. The solution can be obtained by solving Eq. 9.3 in LINGO as follows:

```
!Equation 9.3;
-0.8 * T1 + 0.047 * T2 + 0.003 * T4 + 0.001 * T5 + 0.003 * T6 + 0.002 * T7 +
      0.001 * T8 + 0.002 * T9 + 0.01 * T10 = -0.8;
0.9 * T1 -0.92 * T2 + 0.018 * T3 + 0.025 * T4 + 0.031 * T5 + 0.2 * T6 + 0.014
* T7 +
      0.107 * T8 + 0.023 * T9 + 0.225 * T10 = - 3.0;
0.208 * T1 + 0.692 * T2 -0.7 * T3 + 0.0690 * T4 + 0.0860 * T5 + 0.346 * T6 +
      0.015 * T7 + 0.129 * T8 + 0.008 * T9 + 0.623 * T10 = - 5.81;
0.03 * T1 + 0.004 * T2 - 0.9 * T4 + 0.003 * T5 + 0.03 * T6 + 0.005 * T7
      + 0.027 * T8 + 0.012 * T9 + 0.18 * T10 = - 2.28;
1.08 * T1 + 0.208 * T2 + 0.012 * T3 + 0.08 * T4 - 0.6 * T5 + 0.08 * T6 +
      0.035 * T7 +
      0.053 * T8 + 0.005 * T9 + 0.048 * T10 = - 7.2;
0.15 * T1 + 0.04 * T2 + 0.007 * T3 + 0.05 * T4 + 0.013 * T5 - 0.8 * T6 +
      0.005 * T7 + 0.013 * T8 + 0.003 * T9 + 0.09 * T10 = - 2.4;
1.388 * T1 + 0.37 * T2 + 0.067 * T3 + 0.093 * T4 + 0.347 * T5 + 0.463 * T6 -
0.7 * T7
      + 0.37 * T8 + 0.028 * T9 + 0.389 * T10 = - 2.22;
0.113 * T1 + 0.015 * T2 + 0.005 * T3 + 0.038 * T4 + 0.009 * T5 + 0.038 * T6 +
      0.004 * T7 - 0.97 * T8 + 0.023 * T9 + 0.675 * T10 = - 17.50;
0.495 * T1 + 0.066 * T2 + 0.024 * T3 + 0.165 * T4 + 0.041 * T5 + 0.165 * T6 +
      0.018 * T7 + 0.088 * T8 - 0.85 * T9 + 0.99 * T10 = - 135.68;
0.05 * T1 + 0.033 * T2 + 0.017 * T3 + 0.017 * T4 + 0.017 * T5 + 0.083 * T6 +
      0.013 * T7 + 0.147 * T8 + 0.005 * T9 - 0.88 * T10 = - 4;
```

Note that there is no need for matrix inversion since LINGO can find a solution to the system of equations. A section of the resulting report is shown below:

```
Feasible solution found.
Infeasibilities:                                    0.000000
Total solver iterations:                                   9
Elapsed runtime seconds:                                0.15

Model Class:                                              LP

Total variables:                        11
Nonlinear variables:                     0
Integer variables:                       0

Total constraints:                      11
Nonlinear constraints:                   0

Total nonzeros:                        109
Nonlinear nonzeros:                      0


                                                  Variable          Value
                                                 TOTAL           543.9763

                                                      T1          3.877157
                                                      T2          28.69250
                                                      T3          78.70911
                                                      T4          11.50741
                                                      T5          46.28251
                                                      T6          11.47178
                                                      T7          104.6551
                                                      T8          38.92035
                                                      T9          201.1096
                                                     T10          18.75080
```

The solution report results in a total workforce vector, with figures rounded off to the nearest hundredths, of $\mathbf{t} = (3.88\ 28.69\ 78.71\ 11.51\ 46.28\ 11.47\ 104.66\ 38.92\ 201.11\ 18.75)^T$. As can be seen from the results, even when the changes in external demand were only for medical and nursing staff, all other departments resulted in a decrease in the total workforce needed. The total workforce of the acute care hospital decreases from 563 to 543.98. This clearly shows the interdependency between departments.

9.4.2 Scenario 2

Scenario 2 considers a crisis condition that has resulted in an initial reduction in workforce for all departments as shown in Table 9.6. A 10% reduction was experienced for all departments except for the medical and nursing staff which both experience a higher reduction of 12%. This scenario is similar to that considered by Aviso et al. [1]. Furthermore, each department has to meet exogenously defined levels for the net output of work (**e**). The reduction in the available workforce in each department is represented by the parameter $(1 - \alpha_j)$, while the maximum reduction level in net output of work is given by $(1 - \beta_i)$.

Table 9.6 Limiting assumptions for Scenario 2

Department	Initial available workforce ($\mathbf{t_0}$)	α	Available workforce during crisis ($\mathbf{t}_f$)	Initial net output of work ($\mathbf{e_0}$)	β	Minimum net output of work during crisis ($\mathbf{e^L}$)
1	4	0.90	3.60	0.807	0.75	0.605
2	30	0.90	27.00	2.895	0.75	2.171
3	83	0.90	74.70	5.886	0.75	4.415
4	12	0.90	10.80	2.310	0.75	1.733
5	48	0.90	43.20	7.104	0.75	5.328
6	12	0.90	10.80	2.461	0.75	1.846
7	111	0.90	99.90	2.185	0.75	1.639
8	45	0.88	39.60	22.491	0.85	19.117
9	198	0.88	174.24	30.662	0.85	111.063
10	20	0.90	18.00	3.935	0.60	2.361

The optimization model has the objective of maximizing the total net output of work as given in Eq. 9.5 where $\mathbf{1}^{\mathrm{T}}$ represents a row vector with all elements equal to 1 and $\mathbf{e}_{\text{new}}$ is the optimal net output of work during crisis.

$$\text{Max} = \mathbf{1}^{\mathrm{T}}\mathbf{e}_{\text{new}} \tag{9.5}$$

In the assumption that the interaction matrix remains the same, the optimal solution should meet the conditions of Eqs. 9.6–9.8 where $\mathbf{t}_f$ is the available workforce and $\mathbf{e}_{\text{new}}$ is the net output of work during crisis conditions. The $\boldsymbol{\alpha}^{\mathrm{T}}$ is a row vector which contains the fraction of workforce retained for each department, while $\boldsymbol{\beta}^{\mathrm{T}}$ is a row vector containing the minimum fraction of net output of work that must be delivered by each department during crisis.

$$\mathbf{t}_{\text{new}} \leq \mathbf{t}_{\text{f}} \tag{9.6}$$

$$\mathbf{t}_f = \boldsymbol{\alpha}^{\mathrm{T}}\mathbf{t} \tag{9.7}$$

$$\mathbf{e}_{\text{new}} \geq \boldsymbol{\beta}^{\mathrm{T}}\mathbf{e} \tag{9.8}$$

$$\mathbf{e}_{\text{new}} = (\mathbf{I}-\mathbf{A}) * \mathbf{t}_{\text{new}} \tag{9.9}$$

The solution can be obtained using LINGO as shown below:

```
!Eq. 9.5;
MAX= ENEW1 + ENEW2 + ENEW3 + ENEW4 + ENEW5 + ENEW6 + ENEW7 + ENEW8 +
  ENEW9 + ENEW10;

!Eq. 9.6;
TNEW1 - TF1 <= 0;
TNEW2 - TF2 <= 0;
TNEW3 - TF3 <= 0;
TNEW4 - TF4 <= 0;
TNEW5 - TF5 <= 0;
TNEW6 - TF6 <= 0;
TNEW7 - TF7 <= 0;
TNEW8 - TF8 <= 0;
TNEW9 - TF9 <= 0;
TNEW10 - TF10 <= 0;

!Eq. 9.7;
TF1 = 3.6;
TF2 = 27;
TF3 = 74.7;
TF4 = 10.8;
TF5 = 43.2;
TF6 = 10.8;
TF7 = 99.90000000000001;
TF8 = 39.6;
TF9 = 174.24;
TF10 = 18;

!Eq. 9.8;
ENEW1 >= 0.605;
ENEW2 >= 2.171;
ENEW3 >= 4.415;
ENEW4 >= 1.733;
ENEW5 >= 5.328;
ENEW6 >= 1.846;
ENEW7 >= 1.639;
ENEW8 >= 19.117;
ENEW9 >= 111.063;
ENEW10 >= 2.361;
```

```
!Eq. 9.9;
ENEW1 - 0.8 * TNEW1 + 0.047 * TNEW2 + 0.003 * TNEW4 + 0.001 * TNEW5 +
      0.003 * TNEW6 + 0.002 * TNEW7 + 0.001 * TNEW8 + 0.002 * TNEW9 + 0.01
      * TNEW10 = 0;
0.9 * TNEW1 + ENEW2 - 0.92 * TNEW2 + 0.018 * TNEW3 + 0.025 * TNEW4 + 0.031
      * TNEW5 + 0.2 * TNEW6+ 0.014 * TNEW7 + 0.107 * TNEW8 + 0.023 * TNEW9
      + 0.225 * TNEW10 = 0;
0.208 * TNEW1 + 0.692 * TNEW2 + ENEW3 - 0.7 * TNEW3 + 0.0690 * TNEW4 +
      0.0860 * TNEW5 + 0.346 * TNEW6 + 0.015 * TNEW7 + 0.129 * TNEW8 +
      0.008 * TNEW9 + 0.623 * TNEW10 = 0;
0.03 * TNEW1 + 0.004 * TNEW2 + ENEW4 - 0.9 * TNEW4 + 0.003 * TNEW5 + 0.03
      * TNEW6 + 0.005 * TNEW7 + 0.027 * TNEW8 + 0.012 * TNEW9 + 0.18 *
      TNEW10 = 0;
1.08 * TNEW1 + 0.208 * TNEW2 + 0.012 * TNEW3 + 0.08 * TNEW4 + ENEW5 - 0.6
      * TNEW5 + 0.08 * TNEW6 + 0.035 * TNEW7 + 0.053 * TNEW8 + 0.005 *
      TNEW9 + 0.048 * TNEW10 = 0;
0.15 * TNEW1 + 0.04 * TNEW2 + 0.007 * TNEW3 + 0.05 * TNEW4 + 0.013 * TNEW5
      + ENEW6 - 0.8 * TNEW6 + 0.005 * TNEW7 + 0.013 * TNEW8 + 0.003 *
      TNEW9 + 0.09 * TNEW10 = 0;
1.388 * TNEW1 + 0.37 * TNEW2 + 0.067 * TNEW3 + 0.093 * TNEW4 + 0.347 *
      TNEW5 + 0.463 * TNEW6 + ENEW7 - 0.7 * TNEW7 + 0.37 * TNEW8 + 0.028 *
      TNEW9 + 0.389 * TNEW10 = 0;
0.113 * TNEW1 + 0.015 * TNEW2 + 0.005 * TNEW3 + 0.038 * TNEW4 + 0.009 *
      TNEW5 + 0.038 * TNEW6 + 0.004 * TNEW7 + ENEW8 - 0.97 * TNEW8 + 0.023
      * TNEW9 + 0.675 * TNEW10 = 0;
 0.495 * TNEW1 + 0.066 * TNEW2 + 0.024 * TNEW3 + 0.165 * TNEW4 + 0.041 *
      TNEW5 + 0.165 * TNEW6 + 0.018 * TNEW7 + 0.088 * TNEW8 + ENEW9 - 0.85
      * TNEW9 + 0.99 * TNEW10 = 0;
 0.05 * TNEW1 + 0.033 * TNEW2 + 0.017 * TNEW3 + 0.017 * TNEW4 + 0.017 *
      TNEW5 + 0.083 * TNEW6 + 0.013 * TNEW7 + 0.147 * TNEW8 + 0.005 *
      TNEW9 + ENEW10 - 0.88 * TNEW10 = 0;
```

The solution is summarized in Table 9.7, and it shows that the optimal solution recommends that the workforce needed during crisis in each department may not require 100% utilization to maximize the net output of work of the organization. In this example, the medical staff department (Department 8) only needs 95% of their available workforce to meet their minimum net output of work. The nursing staff (Department 9) on the other hand maximizes the available workforce and is able to function at 89.7% of their full capability if there was no crisis.

Table 9.7 Optimal allocation of workforce between departments in Scenario 2

Department	Available workforce during crisis ($\mathbf{t}_f$)	Workforce needed during crisis ($\mathbf{t}_{new}$)	Minimum net output of work during crisis ($\mathbf{e}^L$)	Net output of work during crisis ($\mathbf{e}_{new}$)
1	3.60	3.28	0.605	0.605
2	27.00	24.93	2.171	2.171
3	74.70	74.70	4.415	9.157
4	10.80	10.80	1.733	2.640
5	43.20	39.55	5.328	5.328
6	10.80	9.82	1.846	1.846
7	99.90	99.90	1.639	6.884
8	39.60	37.64	19.117	19.117
9	174.24	174.24	111.063	117.167
10	18.00	15.90	2.361	2.361

An alternative LINGO model which makes use of the set-based modeling approach is shown below:

```
SETS:

DEPARTMENT: E, EL, ENEW, T, TNEW, ALPHA, BETA, TF;
AMATRIX (DEPARTMENT, DEPARTMENT): IA;

ENDSETS

DATA:

DEPARTMENT = 1..10;

IA =
0.800  -0.047  0.000  -0.003 -0.001 -0.003 -0.002 -0.001 -0.002 -0.010
-0.900 0.920  -0.018  -0.025 -0.031 -0.200 -0.014 -0.107 -0.023 -0.225
-0.208 -0.692 0.700   -0.069 -0.086 -0.346 -0.015 -0.129 -0.008 -0.623
-0.030 -0.004 0.000   0.900  -0.003 -0.030 -0.005 -0.027 -0.012 -0.180
-1.080 -0.208 -0.012  -0.080 0.600  -0.080 -0.035 -0.053 -0.005 -0.048
-0.150 -0.040 -0.007  -0.050 -0.013 0.800  -0.005 -0.013 -0.003 -0.090
-1.388 -0.370 -0.067  -0.093 -0.347 -0.463 0.700  -0.370 -0.028 -0.389
-0.113 -0.015 -0.005  -0.038 -0.009 -0.038 -0.004 0.970  -0.023 -0.675
-0.495 -0.066 -0.024  -0.165 -0.041 -0.165 -0.018 -0.088 0.850  -0.990
-0.050 -0.033 -0.017  -0.017 -0.017 -0.083 -0.013 -0.147 -0.005 0.880
;
T = 4 30    83    12    48    12    111   45    198   20;
BETA   =    0.75  0.75  0.75  0.75  0.75  0.75  0.75  0.85  0.85  0.60;
ALPHA  =    0.90  0.90  0.90  0.90  0.90  0.90  0.90  0.88  0.88  0.90;
ENDDATA
```

```
MAX = @SUM(DEPARTMENT(I): ENEW(I));

@FOR(DEPARTMENT(I): E(I) = @SUM(DEPARTMENT(J): IA(I,J)*T(J)));

!Eq. 9.6;
@FOR(DEPARTMENT(J): TNEW(J) <= TF(J));

!Eq. 9.7;
@FOR(DEPARTMENT(J): TF(J) = ALPHA(J)*T(J));

!Eq. 9.8;
@FOR(DEPARTMENT(I): ENEW(I) >= EL(I));
@FOR(DEPARTMENT (I):EL(I) = BETA(I) *E(I));

!Eq. 9.9;
@FOR(DEPARTMENT(I): ENEW(I) = @SUM(DEPARTMENT(J): IA(I,J)*TNEW(J)));
```

9.5 Conclusions

The input–output modeling framework can also be used to model interactions among groups of workers within an organization. The interactions represent transactions that are measurable in terms of person-hours instead of financial payments. Nevertheless, such transactions can be accounted for via input–output analysis. This methodology can be used to optimally adjust workforce allocation when workload deviates from normal conditions.

References

1. Aviso KB, Mayol AP, Promentilla MAB, Santos JR, Tan RR, Ubando AT, Yu KDS (2018) Allocating human resources in organizations operating under crisis conditions: a fuzzy input-output optimization modeling framework. Resour Conserv Recycl 128:250–258
2. Aviso KB, Cayamanda CD, Mayol AP, Yu KDS (2017) Optimizing human resource allocation in organizations during crisis conditions: a P-graph approach. Process Integr Optim Sustain 1 (1):59–68
3. Correa H, Correa V (1996) An application of input-output analysis to the administration of a library. Libr Inf Sci Res 18(4):343–356
4. Correa H, Craft J (1999) Input–output analysis for organizational human resources management. Omega 27(1):87–99
5. Correa H, Guajardo SA (2001) An application of input–output analysis to a city's municipal government. Socio-Econ Plan Sci 35(2):83–108
6. Correa H, Parker BR (2005) An application of organizational input–output analysis to hospital management. Socio-Econ Plan Sci 39(4):307–333
7. Santos JR, Herrera LC, Yu KDS, Pagsuyoin SAT, Tan RR (2014) State of the art in risk analysis of workforce criticality influencing disaster preparedness for interdependent systems. Risk Anal 34(6):1056–1068

Chapter 10
Future Research Prospects for Input–Output Models

Abstract In this chapter, the potential for further research developments in input–output analysis is discussed. Promising methods (e.g., AHP) for calibrating input–output models are presented to complement mainstream statistical approaches for estimating the coefficients of the model. The prospect of combining input–output analysis with other optimization techniques, such as P-graph and pinch analysis, is also discussed. Finally, this chapter provides reflections on the possible application of extended input–output models to contemporary issues such as climate change or emerging sustainability concepts such as the circular economy.

Keywords Analytic hierarchy process (AHP) · Artificial neural network
P-graph · Pinch analysis · Circular economy

10.1 Introduction

Input–output modeling provides a powerful, versatile, and elegant framework for the analysis and optimization of linear networks. While most applications are used for economic structure analysis, input–output analysis also applies to other networks of analogous structure, such as supply chain networks, ecosystem food chains, and organizational interactions. The preceding chapters have a comprehensive treatment of some of the ways by which input–output models can be used to plan sustainable industrial systems, along with a range of case studies and associated LINGO code for implementation. These examples represent only a small subset of possible applications of the input–output framework. In this final chapter, we discuss potential future directions for research on input–output models.

R. R. Tan et al., *Input–Output Models for Sustainable Industrial Systems*,
Lecture Notes in Management and Industrial Engineering,
https://doi.org/10.1007/978-981-13-1873-3_10

10.2 Model Calibration and Data Acquisition

One of the difficulties in the use of input–output models is that the approach is naturally data-intensive. Coefficients of economic input–output models are derived from records of actual transaction, and the process of compiling and verifying these data requires plenty of time and effort. As a result, input–output models provide a snapshot of the structure of an economy in the recent (or, in some cases, not very recent) past. In developing countries, construction of input–output tables may be subject to greater delays than what is typical in developed nations. As a result, these models are strictly valid only for relatively near-term projections involving negligible technological change. In mainstream input–output analysis literature, there are statistical techniques that are currently in use for updating input–output coefficients. However, alternative approaches are also possible. For example, officially recorded coefficients can be adjusted based on engineering knowledge, as in the case of incremental improvements in the efficiency of existing processes [1]. In the case of new technologies and emerging sectors, it is important to be able to adequately reflect uncertainties within the modeling framework; approaches such as fuzzy set theory have thus been proposed to be used in input–output models, particularly to model imprecise coefficients [2].

A method for estimating input–output coefficients based on the analytic hierarchy process (AHP) was proposed by Saaty and Vargas [3]. This method enables to estimate the matrix of technological coefficients using the ratio scale priority vectors derived from pairwise comparative judgment matrix. Expert judgments can be elicited to determine how strong one sector is as compared to another when allocating outputs to domestic need. A matrix of pairwise comparison among sectors can also be done to describe the sectors' interdependence by eliciting judgment on how strong is the dependence of *i*th sector is as compared with another in receiving output from *j*th sector. Although this approach applies to economic sectors, it can also be extended to socio-technical systems and introduce intangible factors in such resource allocation problem. This approach is potentially useful, either as a stand-alone technique or as a means to augment classical statistical approaches to fill data gaps. Other alternative methods have also been suggested in the literature. For example, the use of artificial neural networks (ANNs) for the dynamic analysis of economic structures was also proposed some time ago [4]. There has been sparse development in this topic, but recent advances in convolutional neural networks and deep learning may be useful for forecasting changes in input–output coefficients.

10.3 Hybrid Modeling Frameworks

Input–output systems are basically a special class of networks. As such, economic structures can also be analyzed from the perspective of network analysis (e.g., [5–7]. Such fusion can bring a larger set of analytical tools to bear on the problem of elucidating economic structures, especially at regional and global scales. These emerging capabilities can provide valuable insights for assessing the sustainability of geographically dispersed economic networks of global scale [8].

In addition to analyzing the current states of economic networks, there is often some interest in using the input–output framework to determine optimal or desirable future states. Such goals can be achieved via different mathematical programming approaches, including those described in this book. Other methods can also be applied to this problem. The process graph or P-graph framework, which was originally developed as a mathematically rigorous and computationally efficient approach to solving process network synthesis (PNS) problems [9–11], makes use of process unit models that have similar assumptions that underlie those of economic sectors in input–output models. In particular, the P-graph framework assumes fixed input–output ratios that result in linear models that are structurally similar to input–output models. This structural similarity was the basis for recent developments in the use of P-graph models to optimize systems having to operate under abnormal conditions due to crisis. Thus far, the approach has been applied to entire economic systems [12] and organizations [13], but can potentially be extended to a wide range of input–output applications, such as system debottlenecking, expansion planning, and system design.

Similarly, the prospect of hybridizing input–output analysis with pinch analysis was recently proposed by Tan et al. [14]. Pinch analysis is a graphical or algebraic technique for determining targets, or benchmarks, for optimum resource use in industrial systems; it was originally developed to optimize heat recovery and energy conservation in process plants [15]. This development is still in its early stages, and further opportunities may eventually become evident. Input–output models also have structural similarities with some decision analysis tools. For example, the popular technique known as DEMATEL (Decision-Making Trial and Evaluation Laboratory) makes use of linear equations resembling input–output models to determine relative influences of various factors in complex decision-making problems [16]. This structural resemblance may suggest further opportunities to extend input–output techniques to other decision analysis applications.

10.4 Input–Output Models and the Future

Input–output models can be very useful for providing a perspective for the optimization of future systems. These models emphasize interdependencies among components of input–output networks, which can in turn give insights for appropriate planning toward sustainability goals. For example, models based on this framework have been applied to the problem of projecting the viability of the commitments made by various countries to the 2015 Paris Climate Accord [1]. Optimization models based on input–output formulation can also be a basis for operationalizing the concept of the circular economy [17], which entails the deployment of new technologies and processes within economic systems to maximize reuse or recycle of waste streams. The provision of a rigorous quantitative framework can support the largely qualitative and conceptual developments in this area.

In his 1986 book *The Blind Watchmaker*, Richard Dawkins wrote that "some of the greatest advances in science have come about because some clever person spotted an analogy between a subject that was already understood, and another still mysterious subject" [18]. This statement underscores the importance of the identification of structural analogies as basis for the transfer of tools and methods from one problem domain to another. The history of input–output analysis, which was originally developed for the analysis of economic systems, illustrates precisely how a versatile and powerful analytical framework can be brought to bear on other systems which, on the surface, appear to be unrelated to the original application. Input analysis has been proven to be applicable to a general class of linear networks that can describe a wide range of systems and phenomena. These systems include some of the topics covered in this book, such as industrial complexes and organizations, as well as other topics beyond its scope, such as food chains and natural ecosystems. Further applications of input–output models and its variants are bound to arise in the future—limited only be the imagination of researchers in this area.

References

1. Cayamanda CD, Aviso KB, Biona JBM, Culaba AB, Promentilla MAB, Tan RR, Ubando AT (2017) Mapping a low-carbon future for the Philippines: scenario results from a fractional programming input–output model. Process Integr Optim Sustain 1:293–299
2. Buckley JJ (1989) Fuzzy input–output analysis. Eur J Oper Res 39:54–60
3. Saaty TL, Vargas LG (1979) Estimating technological coefficients by the analytic hierarchy process. Socio-Econ Plann Sci 13:333–336
4. Wang S (2001) The neural network approach to input-output analysis for economic systems. Neural Comput Appl 10:22–28
5. Blöchl F, Theis FJ, Vega-Redondo F, Fisher EON (2011) Vertex centralities in input–output networks reveal the structure of modern economies. Phys Rev E 83:046127
6. McNerney J, Fath BD, Silverberg G (2013) Network structure of inter-industry flows. Phys A 392:6427–6441
7. Schweitzer F, Fagiolo G, Sornette D, Vega-Redondo F, Vespignani A, White DR (2009) Economic networks: the new challenges. Science 325:422–425
8. Liu X, Klemeš JJ, Varbanov PS, Čuček L, Qian Y (2017) Virtual carbon and water flows embodied in international trade: a review on consumption-based analysis. J Clean Prod 146:20–28
9. Friedler F, Tarjan K, Huang YW, Fan LT (1992) Graph-theoretic approach to process synthesis: axioms and theorems. Chem Eng Sci 47:1973–1988
10. Friedler F, Tarjan K, Huang YW, Fan LT (1992) Combinatorial algorithms for process synthesis. Comput Chem Eng 16:313–320
11. Friedler F, Tarjan K, Huang YW, Fan LT (1993) Graph-theoretic approach to process synthesis: polynomial algorithm for maximal structure generation. Comput Chem Eng 17:929–942

12. Aviso KB, Cayamanda CD, Solis FDB, Danga AMR, Promentilla MAB, Santos JR, Tan RR (2015) P-graph approach for GDP-optimal allocation of resources, commodities and capital in economic systems under climate change-induced crisis conditions. J Clean Prod 92:308–317
13. Aviso KB, Cayamanda CD, Mayol AP, Yu KDS (2017) Optimizing human resource allocation in organizations during crisis conditions: a P-graph approach. Process Integr Optim Sustain 1:59–68
14. Tan RR, Aviso KB, Foo DCY (2018) Carbon emissions pinch analysis of economic systems. J Clean Prod 182:863–871
15. Linnhoff B, Flower JR (1978) Synthesis of heat exchanger networks. AIChE J 24:633–642
16. Gabus A, Fontela E (1972) World problems, an invitation to further thought within the framework of DEMATEL. Battelle Geneva Research Center, Geneva, Switzerland
17. Prieto-Sandoval V, Jaca C, Ormazabal M (2018) Towards a consensus on the circular economy. J Clean Prod 179:605–615
18. Dawkins R (1986) The blind watchmaker. W. W. Norton & Company Inc, NY, USA

Glossary

Constraint An equation or inequality that defines a limitation on the range of values that can be assumed by the variables of an optimization model

Enterprise input-output model A small-scale input-output model whose flows and coefficients are deduced from firm- or enterprise-level data

Environmentally extended input-output model An input-output model augmented with additional equations or inequalities to signify the flows of natural resources drawn from, as well as pollutants discharged into, the environment

Feasible region The region defined by the constraints of a mathematical program which contains all possible values of the system variable

Final demand The net output of a sector in an input-output system that is used for final consumption (i.e., not used as an input for production elsewhere)

Fuzzy optimization model An optimization model whose feasible region is a fuzzy set

Fuzzy set A mathematical set with ill-defined boundaries, such that its elements can assume partial membership

Global optimum A solution of a mathematical program which contains the best possible value of the objective function, as well as the associated values of the variables that lead to this solution

Gross domestic product The total value of the final demand or net output of an economic system, often used as a measure of the size of the system

Industrial complex A network of industrial plants that are linked via a supply chain and are geographically clustered

Infrastructure system A network of basic facilities that provide essential support for activities in society

R. R. Tan et al., *Input–Output Models for Sustainable Industrial Systems*,
Lecture Notes in Management and Industrial Engineering,
https://doi.org/10.1007/978-981-13-1873-3

Inoperability A dimensionless measure of risk in infrastructure systems, whose value ranges from 0 at normal state to 1 at a state of total failure

Inoperability input-output model An input-output model that describes the flow of inoperability risk within an infrastructure, process, or economic network

Input-output model A linear system of equations and/or inequalities that describes a network of interdependent processes, sectors, or activities

Intermediate demand Demand for goods within an economic system, by other sectors that require the goods as production inputs

Leontief inverse The inverse of the difference between an identity matrix and the technical coefficient matrix of an input-output system, which when multiplied by the final demand vector yields the total output vector

Linear program A class of mathematical programming models whose objective function and constraints are comprised entirely of zero- and first-order polynomial terms, and whose variables are all continuous

Life cycle assessment A systematic methodology for the analysis of the environmental impacts of product systems that takes into account cradle-to-grave effects

Local optimum A solution to a mathematical program but which does not necessarily give the best of all possible value of the objective function, but which is superior to solutions in its immediate vicinity

Mathematical program A model consisting of one or more objective functions and a set of constraints that define feasible values of system variables; also known as an *optimization model*

Mixed-integer linear program A class of mathematical programming models whose objective function and constraints are comprised entirely of zero- and first-order polynomial terms, and which contains both continuous and integer variables

Objective function A measure of the performance of a system, expressed as an algebraic function of system variables, which is to be minimized or maximized in an optimization model

Organizational input-output model An input-output model of interactions among groups of staff or personnel, with transactions being measured in person-hours rather than economic value

Perturbation An initial shock or disruption to an input-output system, usually as a result of an external triggering event

Physical input-output model An input-output model whose flowrates and coefficients are given in physical rather than economic terms

Polygeneration The coproduction of electricity with other energy and material outputs within a single, highly efficient facility

Regional input-output model An input-output model which represents a geographically defined subset of a larger economic input-output system

Resilience The capability of a system to recover to its normal state after being subjected to a disruption

Scale invariance A condition that assumes a fixed value for technical coefficients for components of an input-output model, regardless of scale or size

Supply chain A network of linked processes or plants that comprise a sequence of steps in the generation of a product

Sustainability The ability of a system to operate indefinitely while considering economic, social, and environmental aspects

Technical coefficient A parameter that indicates the ratio of production input per unit of output in a component of an input-output system also known as *technological coefficient*

Total output Total level of production by an economic sector in an input-output system, which is sufficient to cover both intermediate and final demand

Index

R. R. Tan et al., *Input–Output Models for Sustainable Industrial Systems*,
Lecture Notes in Management and Industrial Engineering,
https://doi.org/10.1007/978-981-13-1873-3

Printed by Printforce, the Netherlands